Sky and Psyche

Sky and Psyche

The Relationship Between Cosmos and Consciousness

Edited by Nicholas Campion
and Patrick Curry

Published in 2006 by Floris Books
Third printing 2018

British Library CIP Data available
ISBN 978-086315-566-6
Printed by Lightning Source

Contents

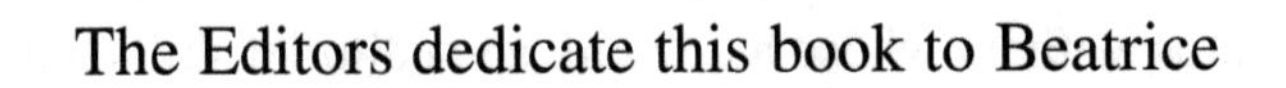

The Editors dedicate this book to Beatrice

Acknowledgments

The Editors would like to thank Alice Ekrek, the Sophia Centre's Administrator, without whose hard work and editorial skills this volume would never have appeared. We would also like to thank the Sophia Trust for their generous funding of the Sophia Centre.

Introduction:
Sky and Psyche — Heaven and Soul

NICHOLAS CAMPION

Director, Sophia Centre, Bath Spa University

The chapters in this book are based on lectures delivered at two conferences held in Bath on May 1 and July 1–2, 2005, respectively: 'The Alchemical Sky' and 'Sky and Psyche.' Both events were initiated by Bath Spa University's Sophia Centre, and were designed, as the title of the second event suggests, to address the question of the relationship between sky and psyche — in their broadest sense. Psyche, in particular, has a double meaning as soul and mind. Until the seventeenth century the two were indistinguishable; soul was that part of mind which could communicate with, travel to, and/or unite with, God. Alchemical sky, meanwhile, points to the possibility of transmutation — or transformation; that the psyche's ability to reflect on the heavens necessarily involves what we might nowadays call an evolutionary process.

The question of the relationship between the soul and the stars has been central to cosmology for thousands of years. The belief in the soul's journey to the stars permeated Egyptian thought. It appeared amongst the Greek Orphics, perhaps under Egyptian influence, from where it made its way into Plato's teachings in fourth-century BCE Athens. Thanks to Plato's impact on the Church Fathers, his theories became a persistent, if controversial, part of Christian theology. In fact, one could argue that the entire Christian notion of soul is pagan. The belief that the soul could embark on a celestial journey draws attention to the cosmos as a real, physical space, one in which morality varies with the region within which one finds one's self. For medieval Christians, Heaven, the soul's natural home, was located above the earth — beyond the stars.

To paraphrase Rob Hand, who spoke at the 'Sky and Psyche' conference, the relationship of soul to stars was *the* central problem in cosmology during the centuries when Christian theology was being formulated and was fighting for supremacy over its pagan and heretical rivals. Did the soul come from the stars? If so, how did it return? Could it return? Did it even want to? These were the sort of questions that pervaded discussions of humanity's relationship with the divine.

While much modern psychology has become almost entirely dissociated from psyche in its original sense, the reaction to such ideas began with Jung in the 1910s and has found a home in the various schools of post-Jungian and transpersonal psychology. Plato's idea of the rational mind, that part of the psyche which was in contact with the divine, survives in various forms. His Idealism, which presented mind as independent of matter, flourishes, unrecognized in a world in which most academic disciplines take materialism as their starting point. Plato's cosmic order though, survives in one other significant area apart from depth psychology, and that is pure mathematics. John Barrow, professor of mathematical sciences at Cambridge University and one of the originators of the anthropic principle, in which the universe and human life are to one degree or another mutually dependent, discussed 'mathematical Platonism,' which he considered 'almost religious in the sense that it provides an underpinning necessary to give meaning to life and human activity.'[1] Roger Penrose is another mathematical Platonist. Professor of Mathematics at Oxford University, he shared the 1988 Wolf Prize for physics with Stephen Hawking. Penrose's explanation for the manner in which mathematics allows for intellectual inquiry is instructive:

> How is it that mathematical ideas can be communicated in this way? I imagine that whenever the mind perceives a mathematical idea, it makes contact with Plato's world of mathematical concepts … when one 'sees' a mathematical truth, one's consciousness breaks through into this world of ideas, and makes direct contact with it ('accessible via the intellect').[2]

Psyche as soul may not survive in Penrose's formula, but psyche as collective mind certainly does, and is not so far from Jung's collective unconscious. Penrose's notion of the individual mind connecting with the world of ideas is certainly an exact replica of the communication that takes place between Plato's human rational soul and the world-soul, the *anima mundi.* Plato laid the foundation of Penrose's opinion in *Phaedrus*:

> Now the divine intelligence, since it is nurtured on mind and pure knowledge, and the intelligence of every soul which is capable of receiving that which befits it, rejoices in seeing reality for a space of time and by gazing upon truth is nourished and made happy until the revolution brings it again to the same place.[3]

Plato's use of the word 'revolution,' of course, is a reference to the revolution of the heavens, of the stars and planets.

To turn to the title of the second, two-day, conference, it deliberately used the word 'psyche' rather than soul; while it is true that psyche is often directly translated as soul, as in English versions of Claudius Ptolemy's *Tetrabiblos*, it is also the root of our modern word psychology — the study of mind. The confusion comes about because, for many in the classical world, God was 'Mind' (*nous* in the Greek), and the human mind, with a small 'm,' was an attribute of each individual's divine consciousness and a means of contacting the Divine. The mind in the modern secular world is viewed, at best, as a set of complexes, at worst as a mere by-product of chemical reactions and electrical impulses, a complicated computer. Many academic and clinical psychologists regard mind as a mere epiphenomenon, an accidental consequence of the brain's physical processes. The word psyche, therefore, in modern terms, deliberately confuses the spiritual and psychological, reminding us that, for much of western history until the modern period, the two were intimately related.

Astronomy, meanwhile, struggles with its origins in celestial religion. When a NASA spokesman describes his reaction to the return of the 'Stardust' mission as an 'incredible thrill, very emotional,' how does this relate to the space programme's overwhelmingly technical logic?[4] When Patrick Moore, the UK's most effective popularizer of astronomy, discussed the 1964 solar eclipse, he simultaneously dismissed ancient beliefs about their power but preserved the notion of the sky as a source of numinous awe:

> Solar eclipses caused great alarm in ancient times; the Chinese used to believe that the Sun was in danger of being eaten by a dragon. No terror is now associated with them, except in very undeveloped countries. But they remain perhaps the most awe-inspiring phenomena in all nature. Nobody who has been fortunate enough to witness a total eclipse of the Sun is ever likely to forget it.[5]

Somehow, astronomy can never quite discard that residual impulse which drew humanity to search the sky for meaning and inspiration.[6] We should turn to Paul Davies, professor of natural philosophy at Adelaide University, for illumination:

> An increasing number of scientists and writers have come to realize that the ability of the physical world to organize itself constitutes a fundamental, and deeply mysterious, property of the universe. The fact that nature has creative power, and is able to produce a progressively richer variety of complex forms and structures, challenges the very foundation of contemporary science. 'The greatest riddle of cosmology,' writes Karl Popper, the well-known philosopher, 'may well be ... that the universe is, in a sense, creative.'[7]

In March 2006, the award of the Templeton Prize to John Barrow again highlighted such prominent opinions on the extent to which the universe is essentially organized and that, therefore, the relationship between consciousness and matter is an integral part of this organization.[8] The relationship between mind and matter may even be purposeful if it is argued that consciousness has developed precisely in order to allow human beings to reflect on the cosmos. News reports of Barrow's award gave renewed prominence to his ideas:

> Life as we know it would be impossible, he and others have pointed out, if certain constants of nature — numbers denoting the relative strengths of fundamental forces and masses of elementary particles — had values much different from the ones they have, leading to the appearance that the universe was 'well tuned for life,' as Dr Barrow put it.
>
> In a news release, the prize organizers said of Dr Barrow's work: 'It has also given theologians and philosophers inescapable questions to consider when examining the very essence of belief, the nature of the universe, and humanity's place in it.'
>
> Asked about his religious beliefs, Dr Barrow said he and his family were members of the United Reformed Church in Cambridge, which teaches 'a traditional deistic picture of the universe,' he said.[9]

Even atheism is no escape from the sky-psyche problem. A recent, ambitious atheist proposal, Frances Crick's 'Astonishing Hypothesis,' singularly fails to provide a reasonable answer. As Crick put it: 'your memories and your ambitions, your sense of personal identity and free will, are in fact no more than the behaviour of a vast assembly of nerve cells and associated molecules.'[10] One might ask if, since the Big Bang, every particle of matter in our bodies has already passed through three stars, including our Sun, and if, as Crick argues, consciousness is a property of matter, at what point in this process does matter develop the ability to inquire into itself? As John Gribbin argued:

> Life begins with the process of star formation. We are made of stardust. Every atom of every element in your body except for hydrogen has been manufactured inside stars, scattered across the universe in great stellar explosions, and recycled to become part of you.[11]

So, to repeat the question, at what stage between star and human do the relevant combinations of Crick's nerve cells and molecules begin to think?

The speakers at the two conferences came from a range of backgrounds. Their brief was to address the topic from whatever was their chosen perspective; personal or professional, academic or practitioner, psychological or spiritual. This variety is reflected in the diversity of the chapters in this book. The intention was not to come up with conclusions but to exchange ideas, for, as none of us know exactly what we mean by soul, or how the mind works, or whether one is a form of the other, the only solution is uncertainty. The universe is a closed system. We are inside it and can never be in the position of impartial, external observers: in reflecting on the cosmos, we are reflecting on ourselves.

Notes

1 Barrow, John, *Pi in the Sky: counting, thinking and being* (London: Penguin, 2002), p. 259.
2 Penrose, Roger, *The Emperor's New Mind: concerning computers, minds and the laws of physics* (London: Vintage, 1991), p. 554.
3 Plato, *Phaedrus*, trans H. N. Fowler (Cambridge Mass., London: Harvard University Press, 1914), 246D, p. 477.
4 BBC Radio 4, 'Today Programme,' 15 January 2006.

5 Moore, Patrick, *Observers Book of Astronomy*, London (Frederick Warne and Co., 1964), p. 158.
6 See the various discussions in Nicholas Campion (ed.), *The Inspiration of Astronomical Phenomena*, Proceedings of the Fourth Conference on the Inspiration of Astronomical Phenomena, sponsored by the Vatican Observatory and the Steward Observatory, Arizona, Magdalen College, Oxford, 3–9 August 2003 (Bristol: Cinnabar Books, 2005).
7 Davies, Paul, *The Cosmic Blueprint: Order and Complexity and the Edge of Chaos*, (London: Penguin, 1995), p. 5, citing Popper, Karl and John Eccles, *The Self and its Brain* (Berlin: Springer International, 1977), p. 61.
8 Barrow, John and Frank Tipler, *The Anthropic Cosmological Principle* (Oxford: Oxford University Press, 1996). For reports and comment on the award of the Templeton Prize to Barrow see Radford, Tim, 'The gods of cosmology,' *The Guardian*, 21 March 2006, p. 33.
9 Overbye, Dennis, 'Math Professor Wins a Coveted Religion Award,' *New York Times,* 16 March 2006, at http://www.nytimes.com/2006/03/16/science/16prize.html?ex=1143176400&en=e587191ce01d41a0&ei=5070&emc=eta1.
10 Crick, Frances, *The Astonishing Hypothesis: the scientific Search for the soul* (London: Simon and Schuster 1994), p. 3.
11 Gribbin, John, *Stardust: the cosmic recycling of stars, planets and people* (London: Penguin 2001), p. 1.

Part I

The Alchemical Sky

1. Love and the Alchemical Saturn

LIZ GREENE

The reasons why people become attracted to, or repelled by, one another are deeply mysterious. Plato calls Eros 'the oldest and most glorious of the gods,' and there is something at work in our attractions and repulsions which can never be entirely psychologized.[1] They feel like fate, or what the Stoics called *Heimarmenê*: the 'compulsion of the stars.'[2] Within families, attractions and repulsions are also deeply mysterious, and ties of blood do not guarantee enduring love or harmony any more than a marriage contract does. Relationships are the chief arena in human life where we experience what we call 'fate,' and even the repudiation of relationship means that it still dominates one's life in the form of strenuous efforts to avoid it. The main body of psychoanalytic literature in the last century has focused primarily on relationship themes — the relationship between conscious and unconscious, and between child and parent — and we have gained many important insights into the kinds of compulsions that drive us into repetitive and sometimes destructive relationship patterns. Countless alternative psychological schools have also offered their understanding, and the arsenal of psychological techniques has grown from Freud's analytic couch to encompass everything from punching pillows to solitary immersion in a tank of tepid water. But no matter how many psychological techniques proliferate, no amount of exploration of childhood as the 'cause' for compulsive relationships can ever plumb that fundamental mystery which focuses on the apparent intelligence at work in 'arranging' our meetings with other humans, and the long-lasting and far-reaching effects these meetings have on our lives. We still do not understand what role fate plays in them; or, indeed, what fate actually is.

Astrologers work with several different approaches when assessing a relationship. Composite charts are valuable maps of a relationship as an autonomous entity, separate from the two individuals comprising it, and expressing the nature of the relationship according to its own psychological laws and necessities.[3] The evaluation of individual birth chart configurations can reveal a person's underlying attitudes toward relationship, especially those linked with parental issues — although parental issues are ultimately only the expression of a profound pattern at work in both the individual and the family matrix, and are never the 'cause' of our attractions and repulsions.[4] Perhaps the most valuable astrological approach to relationship is called *synastry*. This word, like so many astrological terms, comes from the Greek: *syn* (together) and *astron* (a star): literally, a coincidence of stellar influences. Synastry involves comparing two horoscopes in order to see what relationships exist between the planets in one horoscope and the planets in another.[5] In effect, the astrologer working with synastry becomes a fly on the wall, observing an exchange between gods. This very ancient technique remains the chief means by which astrology finds metaphors and meanings for the kinds of responses people have towards each other. Jung's well-known experiment with synchronicity utilized this approach, and involved collecting the charts of couples to look for the ancient signature of the Sun in one chart harmoniously aspecting the Moon in the other — the symbolic 'mystical marriage' or *hierosgamos* reflecting harmony and integration in the heavens, and in the interchange between two human beings.[6]

Astrological synastry has come a long way from the days when your Mars on my Venus meant a sexual compatibility made in heaven, and has become more psychologically sophisticated in interpreting dynamics beyond simple attraction or so-called 'compatibility.' We know, for example, that a Sun-Moon conjunction across two charts, although reflecting the ancient *hierosgamos* in its imagery, may not be quite so comfortable on a human level if the Moon is unknown and unlived in the individual; for we do not always welcome the awakening of an aspect of ourselves which we have been at pains all our lives to suppress or avoid. Nor is 'compatibility,' of the kind described by such a conjunction, always what we seek in relationships, even if we believe it is; perpetual harmony, however pleasant, can also be boring and stagnant. We also know that contacts across two horoscopes involving problematic planets such as that most maligned of heavenly bodies, Saturn, involve complex interactions on many levels, some of which may be exceedingly painful.[7] These contacts may generate an enduring bond, which gives shape and substance to potentials in both people; but they may also lead to the disintegration of a relationship and an internal experience of wounding and bitterness toward the other person and even toward life. And, hopefully, we also know that no matter how revealing the synastry, it still will not tell us whether we 'should' or 'should not' be in a relationship, or how long it will last, or on what level. It might, however, give us hints as to why that relationship has entered our life in the first place, in the sense of a teleology, a meaning, a fate, or perhaps an alchemical work.

The language of astrology describes not only personality characteristics, but also processes of cyclical change and emergence, and it is in this latter domain that it is closely related to the language of alchemy. Alchemical texts are full of astrological references, because the great *opus* was understood to be a work performed on the planetary sympathies within the earth itself — the metals corresponding traditionally to the seven then known planets — with the necessity of aligning the various stages of the work with the relevant celestial patterns to assist a successful outcome. From its earliest days alchemy had a double face: on the one hand the practical chemical work in the laboratory, on the other a psychological process, in part conscious, in part unconsciously projected and experienced through the various transformations of matter.[8] The quest for the incorruptible substance symbolized by alchemical gold was therefore not only literal but also a quest for the creation, or release, of some incorruptible substance within the human being. We may choose any word we like for this incorruptible substance — psychological integration, the Self, the spiritual core, the realization of individual destiny, or, as James Hillman called it, 'the polishing of the maladies,' which is soul-making.[9] The typical or archetypal patterns inherent in the signs, planets, house placements and aspects in an individual horoscope tell us a great deal about those mysterious alchemical processes which occur spontaneously, and without conscious intervention, in every human life; changing us and leading us somewhere, always following the thread of a story, perhaps only dimly glimpsed and perpetually unfinished, but containing its own integrity and its own fate.

The symbolism of alchemy is also applicable to relationship. This should be obvious because alchemical language centres on the image of the *hierosgamos,* the innermost mystery of the art of gold-making. The term most frequently employed for it, *coniunctio*, referred to what we now call chemical combinations, and the substances or 'bodies' to be combined were drawn together by what was called affinity. The word affinity comes from the Latin *affinis*, which means 'bordering on,' that in turn comes from *finis*, an end or a boundary. Affinity is a term applicable in astrological synastry; if I have five planets in Libra, I have affinity with — my boundaries touch or border on — someone who has planets in Gemini or Aquarius, because these all belong to the airy trigon of zodiac signs; and therefore I can expect some degree of empathy and similarity of outlook from such a relationship. Alchemical texts use a variety of terms which all express a human, and more particularly an erotic, relationship, such as *nuptiae, matrimonium, coniugium, attractio, adulatio*. The substances or bodies to be combined were thought of as masculine or feminine, and they were sometimes referred to as dog and bitch, stallion and donkey, cock and hen, and as the winged and wingless dragon. But affinity, in alchemy, is only the beginning. As Jung observed, 'For two personalities to meet is like mixing two chemical substances: if there is any combination at all, both are transformed.'[10] The stages of relationship in alchemy, initiated by the alchemist to facilitate or refine the evolution of nature, are sometimes terrifyingly accurate descriptions of the internal states of relationship dynamics.

Modern astrologers have more heavenly bodies to work with than the alchemists knew about. We have Uranus, Neptune and Pluto, and more recently, Chiron. All these are relevant in comparing two charts and exploring the kinds of psychological responses and transformations that occur when one of these 'substances' combines with a similar or different 'substance' in another human being. It is this issue of the triggering of alchemical processes through relationship dynamics, and what that might look like astrologically, which I would now like to explore through one planet — Saturn — which was written about extensively by the alchemists. The alchemical work begins with Saturn as poisonous lead and ends with Saturn as the reborn polished gold of the Sun. It is, therefore, not surprising that this planet is habitually involved in close contacts across the two birth horoscopes in important relationships.

By 'close contacts' I mean the energetic aspects — conjunctions, squares, oppositions — between Saturn in one horoscope and Saturn, or another planet, in the other horoscope. Saturn is not usually known as a planet associated with *eros*. He is old and cold, grumpy and cantankerous, stern and unbending, and he eats his children. He reflects something within us that feels hurt, deprived, vulnerable, incomplete and determined to compensate through compulsive mastery. But even the most conventionally empathetic and mutually pleasing cross-aspects, such as one person's Sun on another's Moon, or one person's Venus on another's Mars, do not seem to occur as frequently in important relationships, nor do they provide the binding glue which transforms a relationship into an alchemical work. Saturn contacts are also ubiquitous in enduring bonds of friendship, and between couples who have been together for a lifetime, as well as those engaged in brief but intense episodes. Saturn contacts occur in our most lasting and satisfying bonds, and also in those which have been most hurtful to us. They also occur regularly in important business and professional relationships, and between teacher and pupil, guru and disciple. They occur whenever we feel fate at work in a relationship.

We apparently choose our lovers, spouses and friends, although the degree of choice may be questionable in view of the extraordinary way in which people from opposite ends of the earth find each other at the right moment and in the right place. But close Saturn contacts occur consistently within families, and this can hardly be attributed to choice. And Saturn contacts also occur in the charts of those who wish no relationship with each other, but whom an apparently malign fate has thrown together. Although many of you will be unfamiliar with astrological glyphs and symbols, I hope this diagram can visually convey a little of the extraordinary closeness of the Saturn links I am about to describe. Adolf Hitler's Saturn in Leo, which formed a tight square with the Sun and Venus in Taurus in the emperor Hirohito's chart and a Chiron-Neptune conjunction in Josef Stalin's chart, sat firmly on the Sun in Leo in Benito Mussolini's chart, and closely opposed Winston Churchill's Saturn in Aquarius, which in turn closely conjuncted the Aquarian Sun and Venus of Franklin D. Roosevelt (see Figure 1).[11]

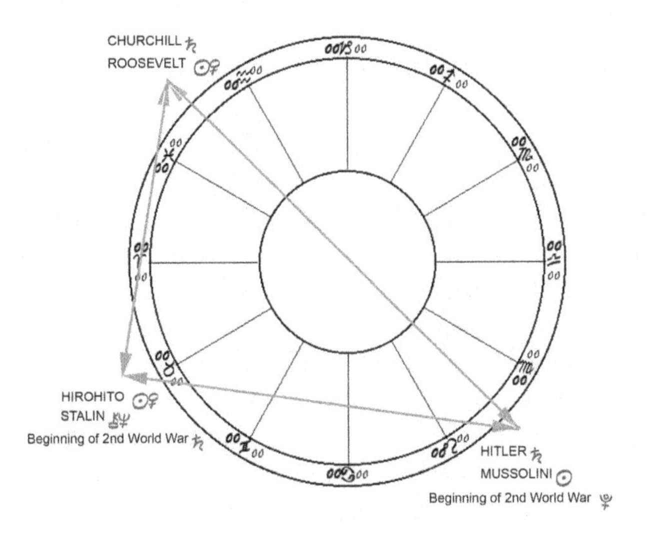

Figure 1: Diagram illustrating Saturn links in the horoscopes of world leaders who were central figures in the Second World War, 1939–45.

These violent collisions of chemical substances drew in millions of lives irrevocably changed, destroyed, scarred, sacrificed, sometimes enobled. Yet these people did not 'cause' the Second World War. They were part of a vast web of planetary connections expressed in part through those whom we call our leaders, who are ultimately no more than transient and usually terrifyingly unconscious mouthpieces for an ongoing process whose beginning vanishes into history, and whose end cannot come as long as there is life. Not surprisingly, at the moment that the war began — when the Germans fired on Danzig — Saturn in the heavens was in Taurus, sitting on Stalin's and Hirohito's planets, setting off the entire configuration; and Pluto in the heavens was in Leo, sitting on Hitler's Saturn and Mussolini's Sun, ensuring that, whatever interaction emerged between these men for good or ill, their personal worlds would be utterly and irrevocably transformed. We still do not fully understand what happened, or what still ongoing alchemical work the collective was subjected to between 1939 and 1945. What we call history is encapsulated in such alchemical collisions, some stunning successes and some horrific failures, and some a mixture of the two.

A national chart may also have powerful Saturn contacts with its leader, elected or otherwise. Once again, I hope this diagram can illustrate something of this strange coincidence of a nation's Saturn so closely and repeatedly linked with its chosen leaders.

America's Saturn at 15° Libra closely conjuncts George Bush's Chiron, Moon and Jupiter in Libra. And he is not the only political figure to constellate the embrace of America's Saturn: Bill Clinton has five planets and the Ascendant all in Libra, all on that American Saturn; John Kennedy's Libran Ascendant fell on that American Saturn; even Senator Joe McCarthy had Venus and Mars in Libra on that American Saturn. All these men — and many other American political leaders — precipitated soul-searching and upheaval in their national collective; some through their own questionable behaviour and some, like Kennedy, through what 'happened' to them as a kind of fate. Is this passion? Perhaps it is a form of *eros* when a nation chooses, follows, allows or colludes with a particular leader. And a leader can precipitate an alchemical process in the body of a nation that is no different from the transformations that occur between two people when Saturn in one of the charts is brought alive through the relationship. Alchemy on such a vast scale suggests that our personal concerns, important and valid though they may be, are

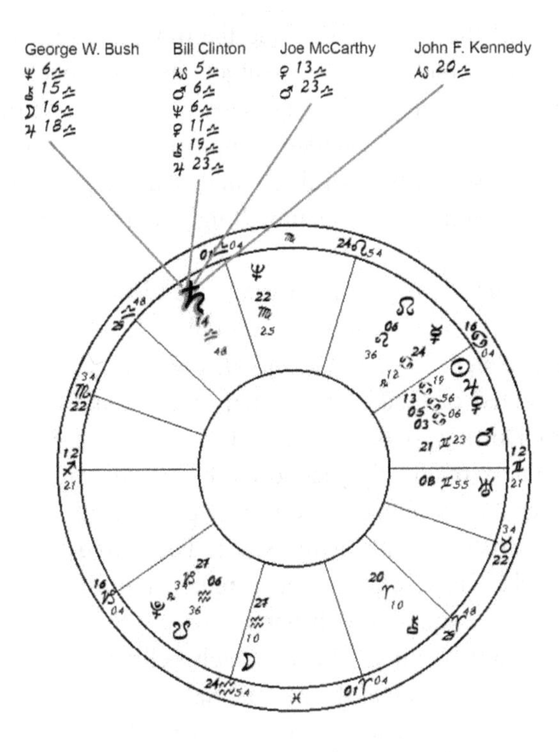

Figure 2: Horoscope of the USA and the planetary conjunctions of some of the key figures of US History in the last century with the USA's Saturn.

tiny alchemical works within greater works and greater alembics, within a vast life of which we have only the dimmest inkling.

Saturn contacts can feel binding, claustrophobic, and inescapable. They may begin with fascination and even passion, but they always change into something else, revealing over time a secret face which points to a deeper and more complex meaning. They force us to become embodied, to feel our heaviness, inadequacies, durability, aloneness and our mortality. They may reflect loyal support or malignant possessiveness, corrosive envy or sincere and respectful admiration. Or we may feel the other as a burden; they, rather than our own selves, seem to become the prison walls beyond which we cannot pass, and which curtail the limitless possibilities we seek. Power issues are usually involved with Saturn contacts, as is the mobilization of defences, for Saturn is the original inventor of the siege engine and the moat and drawbridge. But these bonds take us somewhere. They subject us to a process the results of which may not be visible for many years or, in the case of a nation, many generations. We are chiselled, chipped, riven, sanded, honed, filled and polished; sometimes willingly and with deep satisfaction and a sense of achievement; sometimes unwillingly and with great pain; and sometimes both.

The *prima materia* is the basis of the alchemical *opus*. We have no idea what it really was as a physical substance. It is called Nile mud, red earth, excrement, lead, vinegar, blood, poison, shadow and chaos; sometimes 'accursed of God,' 'the animal of earth and sea,' or simply 'man.' Gerhard Dorn, a pupil of Paracelsus, calls the *prima materia Adamica*: the black, magically fecund earth that Adam took with him from Paradise, or Adam himself, the prototype formed of red clay into which God infused the breath of the spirit. The *prima materia* is also imaged as the Old King, impotent or crippled like the Fisher King in the Grail legend, and equated with Saturn, whose golden solar child is concealed in his dark depths.[12] Lead, which belongs to Saturn, is the heaviest and most poisonous of the metals. Saturn is thus a metaphor for the raw, undifferentiated substance in both the world and the individual, containing the secret of what we most urgently need to develop in order to make something authentic of ourselves. Saturn is our capacity to give shape to what is within us, and our relationship with the laws and limits of the natural world. He is a dark and heavy god, defensive and separative, and he usually drags behind him a sack full of parental hurts, childhood deprivations, early frustrations, and a gnawing sense of inferiority or deformity. These early frustrations are not 'caused' by malign or incompetent parents. They arise because of the gap between those two ordinary flawed humans and the vision of the alchemical stone we have been seeking from birth, encased within Saturn's shell. Because of this gap, we are often full of envy, resentment and rage, which is sometimes easier to bear than the lonely task of struggling to give shape to the incorruptible thing which lies at the core of one's life. Saturn as *prima materia* must be subjected to processes which strip it of its dross and release its secret essence. It is these processes which are set in motion when a close contact involving Saturn occurs between the charts of two people in relationship.

The alembic in which the *opus* is performed must be strong enough to resist extreme heat and the violent transformations occurring within it. It must be sealed during the work so that the living beings envisioned as inhabiting the substances cannot escape. It must also, we are told, be round in imitation of the spherical cosmos, so that the influence of the heavens may contribute to the success of the operation. Chemical processes do not happen at random, but only at the astrologically right moment; and *kairos*, the 'right moment,' is as fundamental to alchemy as it is to astrology. It is also fundamental to relationships, which enter our lives in accord with precisely the right transits occurring across our birth horoscope — not 'because' lumps of rock in the sky are making it so, but, as a child might say, just because. In everyday life the alembic may be a marriage, or a blood tie within a family; it may take the form of work obligations which force contact with particular individuals; it may be financial bonds which endure long after the love has grown cold; it may be consciously constructed in a relationship such as psychotherapy or an astrological chart reading; or it may arise from the sealed chamber of one's own internal needs, compulsions and dependencies. Whatever its form, it must be strong enough to ensure that we remain within the relationship at least long enough to undergo the commencement of the necessary processes, even if the work is unfinished when the relationship ends.

The substances or 'bodies' in the alembic are then subjected to the stages of the *opus*. Central to these stages is the *nigredo*, the blackening. We are told by alchemical texts that 'Saturn reigns in the *nigredo*.'[13] It is a process of cooking the *prima materia* — in other words, cooking Saturn — to break the substance down into its fundamental components, its bare and indestructible bones; washing or burning away the dross, the inauthentic and the contrived, and revealing the indestructible essence. The *nigredo* is something which life brings upon us: any experience which forces us to discover what endures in us when all else has been taken away. Loss, separation, sickness, loneliness, biological changes such as puberty and menopause, and confrontation with collective forces greater than ourselves can break the personality down into its basic components, releasing parental and familial complexes and shattering the fragile defence mechanisms which have previously supported the functioning of the conscious ego. Even 'happy' experiences such as falling in love or bearing a child can precipitate this process, because an old self must die in order to make way for anything new. Life thus presents us regularly, and perhaps cyclically in harmony with Saturn's transit cycle, with a spontaneous *nigredo*, a process which the alchemists observed occurred regularly in Nature in its most fundamental seasonal cycles, and in the decomposition of all dead organic forms. *Nigredo* is also

Figure 3: The Old King (Saturn) calling out for help as he drowns. From Michael Maier's Atalanta Fugiens *(1618).*

one of the most characteristic emotional consequences of the complex mutual reactions decribed by close Saturn contacts across two charts.

I would like to briefly show you some of the images of Saturn's *nigredo* portrayed in alchemical texts. These can describe with sometimes disturbing accuracy the ways in which we are cooked when we become deeply involved with individuals whose planets combine with our natal Saturn.

Saturn can be dissolved. Figure 3 is from Michael Maier's *Atalanta fugiens* (1618), and presents us with the Old King calling out for help as he drowns. He is overwhelmed by the waters, broken down into his essential components through a cleansing in the Great Deep. As a reflection of Saturn in relationship, this image can tell us something about the sense of being overwhelmed by emotions over which we have no control. Sometimes this is the 'two hearts beating as one' experience, a loss of boundaries which at the beginning of an intense relationship seems erotically ecstatic, but which soon evokes fear and defensiveness. Or the drowning may be a crippling feeling of dependency that dissolves the ego's pride and self-sufficiency. It may also be the unwelcome eruption of long-forgotten hurts, and nightmares of abandonment and extinction. This image speaks of being inundated, unable to breathe because the emotional affects are too strong.

The *solutio*, as it is called in alchemy, is potentially a cleansing and profoundly healing process. But Saturn may desperately attempt to regain control by controlling the other person, diminishing them so that they do not seem so powerful. This is the interaction so frequently described by astrological textbooks: the Saturn person is 'overly critical' or 'restrictive' or 'cold.' The vulnerability of dissolution can be terrifying. Yet if we can allow ourselves to be immersed, and suffer the anxiety rather than attempting to diminish the person who has triggered the tidal wave, the Old King may ultimately be renewed, transformed through authenticity and the humility of admitting need.

Saturn can also be boiled. Figure 4 is from Salomon Trismosin's *Splendor Solis* (1582), and it is once again an image of water. But this is water heated to the point where the essence of the *prima materia* is steamed out as an essence. The essence takes the form of a white bird: the Christian symbol of the Holy Spirit, released from the primal substance. One may wonder what the Holy Spirit is doing embedded in Saturn in the first place: a rather heretical idea, since this image informs

Figure 4: The Old King being boiled in water. From Salomon Trismosin's Splendor Solis *(1582).*

Figure 5: The Old King (Saturn) being cooked in a oven. From Michael Maier's Atalanta Fugiens *(1618).*

us that divinity does not come from heaven but is inherent, not only in our bodies but also in the most frightened and defensive sides of the personality.

The Old King must be boiled down to bone, the enduring components of the bare structure of character, freed of the accumulation of false muscles and false flesh. What is it we most truly need to develop in order to feel authentic? Saturn in the birth horoscope is linked with feelings of deprivation and lack we may not even realize we carry. It may take another person to make us aware, for we don't envy qualities in others that mean nothing to us; nor would we would seek to steal them from others if we did not need them to become real. The secret hidden in Saturn's boiled bones is the thing which can give us solid substance, and it is not found in heaven or any attempt to transcend Saturn's earthly world. And it may be steamed out of us in relationship, when our feelings are heated in a pressure cooker but are never allowed to explode or escape. We may bubble but never boil over, cooking until we can distinguish between what we actually are from what we have always believed ourselves to be. It is also interesting to note that the white bird is the Holy Spirit only in Christian doctrine; much earlier, it belonged to Venus, Aphrodite, Ishtar: all the great goddesses of fecundity and erotic love. The essence hidden in the bones may permit a renewed relationship with the body, and the beauty and pleasure of the things of this world.

Saturn can also be roasted like a Sunday side of beef. Figure 5, also from Michael Maier's *Atalanta fugiens*, presents us with the Old King in an oven, with only a tiny candle to illuminate the darkness, while the

Figure 6: The Old King (Saturn) being eaten by a wolf. From Michael Maier's Atalanta Fugiens *(1618).*

fire beneath him sweats the black bile from his body. This might be seen as an image of enforced frustration in relationship, rage unable to be expressed or enacted; the bitterness is slowly sweated out of us because there is nowhere else for it to go. We may burn for revenge or hate as much as we love, but our bond locks us within our own psychic prison: we cannot destroy what we love. Yet there is enough light from the candle to see that, however great the frustration, it goes somewhere and does something, and it may ultimately purge poisons accrued from a lifetime of thwarted expectations and unfulfilled dreams.

Michael Maier also shows us in *Atalanta fugiens* that Saturn can be eaten by a wolf (see Figure 6). Then, having digested the Old King, the wolf is itself sacrificed in the fire, and the King emerges from its ashes reborn. What does it feel like to be eaten alive by a wolf who is then burned in the fire? The wolf is perpetually hungry — the loner, the outsider, the outlaw — it is the animal of Mars, brave but insatiable, forever on the prowl. This is a state of being consumed by hunger for a fantasy, attempting to devour the desired one but then discovering that the person we have tried to consume can never love us enough in return to satiate our craving for something ultimately unobtainable from any other living being. Activated in this way in a relationship, Saturn hungers, possesses and is then himself devoured, coerced into relinquishing what he must sooner or later find within himself. What may rise from the ashes is self-ownership, which is perhaps another way of describing integrity.

Finally, Saturn can be dismembered and then reconstructed from the pieces of his own corpse and lifted from his coffin alive and whole (see Figure 7). Dismemberment, like drowning, boiling, roasting and being eaten alive, is a mythic image. Many gods are dismembered: Osiris is torn to pieces by Set, Dionysus is ripped apart by the Titans, Orpheus has his head clawed from his body by maenads. Dismemberment can be an image of insoluble conflict, a sense of being torn in half, fragmented, lost: an internal battle that seems to allow no resolution, no 'right' choice, no means of retaining one's integrity and one's commitment. The burial and the long wait in the darkness are also mythic images. The claustrophobia of the duties, obligations and responsibilities invoked by the worldly paraphernalia of so many of our emotional bonds — marriage, children, mortgages, joint bank accounts, divorce settlements, maintenance payments, the care of elderly parents — can indeed feel like the suffocating, airless darkness of a tomb in which we wait without hope, not knowing what point there might be in such imprisonment. If we are winged creatures, eternal youths seeking the currents of the air, it can be especially painful and diminishing. Yet in the alchemical image, Saturn arises renewed, the pieces miraculously joined again, one incorruptible body, like the resurrected on Judgement Day.

Acted out in relationship, Saturn's *nigredo* can be a metaphor for the disintegration of our cherished self-image, at the mercy of a process

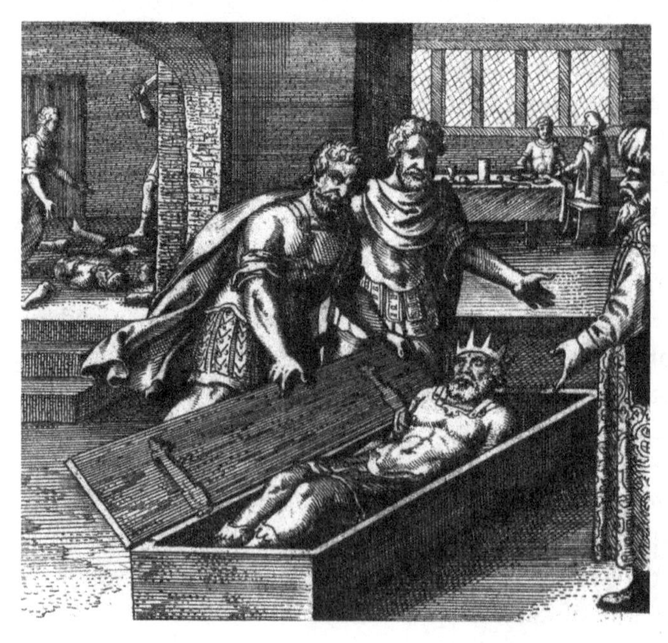

Figure 7: The Old King (Saturn) being lifted from his coffin, alive and whole. From Michael Maier's Atalanta Fugiens *(1618).*

which leads us into becoming what we have always needed to be. Of course, it isn't always as dramatic as being eaten by a wolf or buried alive. Saturn contacts across two charts may cook us through humbler, although equally primal, experiences: for example, our capacity for meanness and petty spite, our anal retentiveness, our remarkable ability to neglect the word of support just when it is most needed, or offer the word of criticism just when it is most hurtful. Saturn can be deeply anal; he withholds, grasps, pinches pennies, complains about the state of the world, and reminds us constantly that no one is as hard done by as he is. He is the eternal whiner of the planetary pantheon, and the looming face of our old age and eventual extinction. Yet if we are prepared to embrace this sour god who once ruled a Golden Age and cannot ever cease reminding us of it, it could mark the first stage of an initiatory journey which will not help us to transcend life, but may help us to become more genuinely earthed in it: someone real, who does not depend on stolen archetypal powers or other people's affirmations in order to endure and build.

It is not only the individual's Saturn that is drawn into this process. If two chemical substances meet, both are changed, and the other planet in the other person's horoscope also suffers, awakens and transforms. That planet is spurred into development by the very fact that it is thwarted by Saturn's defences; for nothing is as powerful a spur to expression as another person telling us, overtly or covertly, that we must not be something we didn't even realize we needed to become. The 'affinity' between Saturn and that other planet is the affinity of possession and the rebellion against it; rejection and the urgency to be wanted; denial and the demand to have. Often the conflict seems irreconcilable, but asking how to solve it in a relationship is probably the wrong question. Turning one's gaze away from the person causing all the trouble, and looking with greater concentration at the bile being sweated out of oneself inside an oven lit by a small candle, may, to borrow James Hillman's term, be more *efficacious*. Whatever 'happens' in the relationship, something crystallizes, hardens, coagulates within oneself: something solid is formed. Whether Saturnian relationships end or continue, in whatever form, they can be the catalysts for a gradual renewal of the personality — through the struggle to integrate qualities previously projected on the loved one, the formation of a more authentic personality, and sometimes the simple discovery that one has survived something profoundly distressing and has emerged still whole and still able to love.

I would like to end by voicing a question which inevitably arises out of this brief exploration of the astrological Saturn in relationship.

In the alchemical work, the substances are chosen and placed in the alembic according to the alchemist's intelligence and intent; and the inauguration of the *nigredo* through fire or water is timed by the alchemist according to the presiding configurations of the heavens. But if we ourselves are the raw substances on which the work is performed, what mysterious intelligence has placed us in the alembic which we call the limitations of our everyday lives, and has chosen which substances to mix together, and has elected the time when we meet those who enter our lives to change us for good or ill? What mysterious intelligence raises up leaders at the moment when their nations have unconsciously invoked them, or conjures enemies when we most need to be subjected to the fire of their envy and hatred? Our meetings seem so much like chance, yet we know that they are not. We may, perhaps, choose to remain in a difficult relationship and allow ourselves to be cooked; or we may elect to smash the alembic from within and break free — although even that act does not prevent the process from being set in motion, and may be the result of a process already undergone. If alchemist and substances are secretly the same, as alchemical texts tell us, then this mysterious intelligence is within us as well as around us, and not some phenomenon 'caused by' the circling of chunks of rock around a dying star. If we call it fate, then we must acknowledge that fate is not 'written in the stars' but is a property of the substance of life itself. It is awesome to contemplate such an intelligence, which transcends time and space and brings us to the right person at precisely the right moment, with precisely the right transiting planets and precisely the right Saturn aspects across our natal charts. Perhaps that intelligence is what Paracelsus described as *magia*: the essence of both the Creator and creation. Or perhaps our relationships are really omens, portents, signatures of the will of the gods as authoritative and precise as any Delphic oracle, if only we could read them. Perhaps our instinctive need, and capacity, to recognize these signatures by which to divine who we are meant to be is best served, not only by scrutinizing the heavens, or by scrutinizing our childhood deprivations, but also by seeing with different eyes, as though in a mirror, those individuals with whom we are locked in resentment, envy and hatred; and even more, those individuals with whom we fall deeply in love.

Notes

1 Plato, *Symposium,* 180b, in *The Collected Dialogues of Plato*, Edith Hamilton and Huntington Cairns, eds. (Princeton: Princeton University Press, 1961).
2 Jung, C. G., *The Spirit in Man, Art and Literature* (Princeton: Princeton University Press, 1966), para. 31.
3 On composite charts see Hand, Robert, *Planets in Composite: Analyzing Human Relationships* (Gloucester, MA: Para Research, 1975); Greene, Liz, *Relationships and How to Survive Them* (London: CPA Press, 1999 [2004]).
4 On parental complexes in the birth chart see Sharman-Burke, Juliet, *The Family Inheritance* (London: CPA Press, 1996); Greene, Liz, 'The Parental Marriage in the Horoscope,' in Greene, Liz and Howard Sasportas, *The Development of the Personality* (York Beach, ME: Samuel Weiser, Inc., 1987).
5 On synastry see Neville, E. W., *Planets in Synastry* (West Chester, PA: Whitford Press, 1990); Thornton, Penny, *Synastry* (Wellingborough, Northamptonshire: Aquarian Press, 1982).
6 Jung, C. G., 'Synchronicity,' in *The Structure and Dynamics of the Psyche, CW8* (London: Routledge & Kegan Paul, 1960), pp. 417–532.
7 For Saturn in synastry, see Greene, Liz, *Saturn: A New Look at an Old Devil* (York Beach, ME: Samuel Weiser, Inc., 1976).
8 On the psychology of alchemical symbolism, see Jung, C. G., *Psychology and Alchemy*, *CW12* (London: Routledge, 1953); Von Franz, Marie-Louise, *Alchemy: An Introduction to the Symbolism and the Psychology* (Toronto: Inner City Books, 1980).
9 Hillman, James, 'Heaven Retains Within its Spheres Half of all Bodies and Maladies (Paracelsus): Astrology as Cosmology' (paper presented at the *Alchemical Sky* conference, 1 May 2005, Bath, UK).
10 Jung, C. G., *The Practice of Psychotherapy*, *CW16* (London: Routledge, 1954), para. 163.
11 All birth data from Taeger, Hans Hinrich, *Internationales Horoskope Lexikon* (Freiburg im Breisgau: Bauer, 1991).
12 See Jung, Emma and Marie-Louise Von Franz, *The Grail Legend* (London: Hodder & Stoughton, 1971).
13 Maier, Michael, *Symbola aureae mensae duodecim nationum* (Frankfurt: 1617), quoted in Jung, C. G., *Mysterium Coniunctionis*, *CW*14 (London: Routledge, 1963), p. 229, fn. 585.

2. The Azure Vault: The Caelum as Experience

JAMES HILLMAN

I would like to begin along two parallel paths: a poem by Lisel Mueller called 'Monet Refuses the Operation' — a cataract operation — and a moment in the analysis of Anna O, reported by Josef Breuer.

So that you may have a general idea of where we are going, I'll state again the title of this talk: 'The Azure Vault: *Caelum* as Experience.' Briefly, the Latin word means the blue sky; heaven; the abode of the gods and the gods collectively; the sky as the breath of life, the air; and also the upper firmament or covering dome, including the Zodiac. The alchemical *caelum,* or *coelum,* is expanded upon especially in Jung's last great work, *Mysterium Coniunctionis*, and, as he says, the *caelum* has 'a thousand names.' These few will also help us as we proceed: 'a Heavenly Spirit that makes its way into the essential forms of things:'[1] the '*anima mundi* in matter;' 'the truth itself;' 'a universal medicine;' 'a window into eternity;' radiating 'a magic power;' 'the *unus mundus*' as a '*unio mystica* with the potential world, or *mundus archetypus*' and the final realization of the alchemical opus.[2] We are headed to the edge.

Now to the poem, and the first of many stories:

'Monet Refuses the Operation'

Doctor, you say there are no haloes
around the streetlights in Paris
and what I see is an aberration
caused by old age, an affliction,
I tell you it has taken me all my life
to arrive at the vision of gas lamps as angels,
to soften and blur and finally banish
the edges you regret I don't see,
to learn that the line I called the horizon

does not exist and sky and water,
so long apart, are the same state of being.
Fifty-four years before I could see
Rouen cathedral is built
of parallel shafts of sun,
and now you want to restore
my youthful errors: fixed
notions of top and bottom,
the illusion of three-dimensional space,
wisteria separate
from the bridge it covers.

What can I say to convince you
The Houses of Parliament dissolves
Night after night to become
The fluid dream of the Thames?
I will not return to a universe
of objects that don't know each other,
as if islands were not the lost children
of one great continent. The world
is flux, and light becomes what it touches,
becomes water, lilies on water,
above and below water,
becomes lilac and mauve and yellow
and white and cerulean lamps,
small fists passing sunlight
so quickly to one another
that it would take long, streaming hair inside my brush to
  catch it.
To paint the speed of light!
Our weighted shapes, these verticals,
burn to mix with air
and change our bones, skin, clothes
to gasses. Doctor,
if only you could see
how heaven pulls earth into its arms
and how infinitely the heart expands
to claim this world, blue vapor without end.

Now the story of Anna O, as told by Josef Breuer: 'She told me there was something the matter with her eyes; she was seeing colours wrong. She knew she was wearing a brown dress but she saw it as a blue one.'[3] Breuer tested her colour vision; it was not impaired. So, Breuer interprets this curious 'misperception' as an incursion of a secondary state of mental functioning into 'her first, more normal one.' Breuer writes: 'She had been very busy with a dressing-gown for her father, which was made with the same material as her present dress, but was blue instead of brown.' The visual mistake — or visionary experience? — is reduced by Breuer to the blue material of the father's dressing gown.

Could there be something more? Could the patient's own avowal that she saw blue despite knowing 'she was seeing colours wrong' indicate a wish, not for the father only, and for dressing his body, but for clothing herself in blue? And, what might it imply for her, for that analysis, for the field of analysis itself — since Anna O is the *fons et origo* of our heritage — for the body of the patient, of every analytical patient, the *opus* itself to be clothed in blue?

We shall return to both Monet and Anna O, but, first, a few more blue stories. Again, one from the beginnings of our field, the devastating crisis starting at age thirty-nine in the life and work of Gustav Theodor Fechner (1801–87), to whom both Sigmund Freud and William James pay effusive tribute, claiming him to be the most valuable thinker in psychology of the nineteenth century. Fechner was a brilliant physical psychologist, observer, micro-measurer, laboratory experimenter.

Then his eyes gave out. He couldn't observe, he couldn't read. He was not blind, but he could no longer see. Nor could he eat or drink. His digestion gone, shrivelled, despairing, sleepless and silent, he protected his eyes with lead cups and retreated into a blackened room, kept alive by his wife. After close to three years in this black hole, gradually recovering physically, he emerged, lifted the bandages and allowed light into his eyes:

> I stepped out for the first time from my darkened chamber and into the garden. ... It seemed to me like a glimpse beyond the boundary of human experience. Every flower beamed upon me with a peculiar clarity, as though into the outer light it was casting a light of its own. To me the whole garden seemed transfigured, as though it were not I but nature that had just arisen. And I thought: So nothing is needed but to open the eyes afresh.

> The picture of the garden accompanied me into the darkened chamber; but in the dusk it was all the brighter and clearer and more beautiful, and at once I thought I saw an inward light as the source of the outward clarity ... and the shining of the plants' souls.[4]

Fechner felt giddy with joy of this moment of the *multi flores*. Fechner went on to live healthily to 86 years of age, immensely productive, exchanging his professorship from natural science to the philosophy of nature. The first lectures he delivered after his recovery were devoted to pleasure, formulating for psychology the pleasure principle later expanded upon by Freud.[5] The book on the soul that followed his return to life was subtitled 'a walk through the visible world in order to find the invisible.'

Fechner now wore blue glasses: to protect his eyes? Or to protect his vision from the materialist perspective that had led to his blindness and which he now called the 'nightworld;' that is, the *nigredo* from which he had emerged.

Two stories from childhood, one from an Irish poet, the other from an American musician. AE (George William Russell) friend of Yeats and pivotal figure in the Irish literary revival, describes a moment when, at about four or five years old, he was lying flat on some grass recalling a story of:

> … a magic sword with a hilt of silver and a blade of blue steel. The word 'magic' stirred me, though I knew not what it meant. … It lay in memory … until a dozen years later its transcendental significance emerged as a glittering dragonfly might come out of a dull chrysalis. The harmony of blue and silver at once bewitched me. I murmured to myself "blue and silver! Blue and silver!" And then, the love of colour awakened … one colour after another entered the imagination. … This love of colour seemed instinctive in the outer nature.

Here, says AE, is 'the birth of the aesthetic sense,' in harmony with the natural world and its colours that were, as he says, 'of its nature and not of that unthinking child's.'[6]

The second story is from the biography of Miles Davis:

> The very first thing I remember in my early childhood says the incomparable artist Miles Davis, a flame, a blue flame

> jumping off a gas stove. ... I remember being shocked ... by the suddenness of it. ... that stove flame is as clear as music is in my mind. I was three years old.
>
> I saw that flame and felt that hotness of it close to my face. I felt fear, real fear ... But I remember it also like some kind of weird joy, too. I guess that experience took me someplace in my head I hadn't been before. To some frontier, the edge, maybe, of everything possible. ... The fear I had was almost like an invitation, a challenge to go forward into something I knew nothing about ... everything I believe in started with that moment. I have always believed and thought since then that my motion had to be forward, away from the heat of that flame.[7]

'Away from the heat of that flame' and into the 'cool;' his inventive use of the mute, his solos as 'thinking' the music, the titles of his great pieces such as 'Kind of Blue,' 'Blue in Green,' his wearing 'shades' already in the 1940s. I can imagine that both Goethe and Kandinsky would approve of Miles' feeling for blue. 'Blue,' says Goethe, 'gives us the impression of cold ... and reminds us of shade. A Blue surface seems to recede from us ... it draws us after it.' Kandinsky adds '[B]lue ... retreat from the spectator ... turning in upon its own centre ... active coolness.' For Davis, a challenge to go forward; for AE, the sympathy for revolt.

The call of AE's aesthetic sense and the spirit of revolt marked the Romantic impulse of which Novalis's 'blue flower' is the undying example — although he himself, as Friederich von Hardenberg, died at the age of twenty-nine. That famous blue flower appears in Novalis's novel of the poetic education in which the hero, Heinrich, dreams a vision. Climbing through a strange geography, the hero comes to a place filled with 'a holy stillness,' where 'a basin of water emits a faint blue light.'[8] Blue, veined cliff ... The sky was blue, clear, and he was 'drawn to a tall light-blue flower. The flower then leaned towards him and ... upon a great blue corolla, hovered a delicate face.'[9]

Novalis regarded the blue flower as 'the visible spirit of song,' (perhaps a revelation of Leibniz's cosmological idea of a pre-established harmony governing all things?) Curious that AE entitles the little book describing the evolution of his poetic calling, *Song and its Fountain*. Novalis writes: 'one thing recalls all ...'[10] 'No more the order of time and space ... the great soul of the world moves everywhere, blooms ceaselessly ... World becomes dream, dream becomes world.'[11]

Another report from the history of our field: this time, Wilhelm Reich. Reich combined the libido of Breuer and Freud with the physical science of Fechner. Reich imagined that the libidinal charge flowing through the body is the "orgone" energy of the cosmos. Freud's later theory of Eros as a cosmic force Reich would capture in a box in which a patient could receive "orgone" radiation. The radiation, according to Reich, came in three variations of blue.[12] Whether Reich was a crackpot or a brilliant therapist does not concern us here: but his witness to blue as the colour of libidinal Eros that embraces the phenomenal world adds another page to our collection of stories. Besides, why not imagine libidinal desire as blue? Weren't porn shows once called 'blue movies' and the suppression of libido attributed to puritans named 'blue stockings'?

Backing for Reich's blue Eros as the universal energy that joins phenomena together comes from Cézanne. I quote from one of his most astute and studious biographers:

> Cézanne gave blue a new depth of meaning ... by making it the foundation of the world of objects 'existing together.' Blue was now recognized as belonging to a deeper level of existence. It expressed the essence of things and their abiding, inherent permanence ... [13]

Cézanne himself wrote: 'Blue gives other colours their vibration, so one must bring a certain amount of blue into a painting.'[14] Zola, referring indirectly to Cézanne, writes: 'the flesh colours are blue, the trees are blue, surely he went over the whole picture with blue.'[15] In his old age Cézanne drew with a brush loaded with aquamarine. May we say Cézanne painted with the color of the *caelum* in order to present the *unus mundus*?

Again from the history of our field, two stories from Jung's biography. In 1944 Jung suffered a heart attack.

> I experienced dreams and visions which must have begun when I hung on the edge of death ... I had reached the outermost limit ... It seemed to me that I was high up in space. Far below I saw the globe of the earth, bathed in a gloriously blue light. I saw the deep blue sea and the continents ... its global shape shone with a silvery gleam through the wonderful blue light ...[16]

Jung's vision goes on for pages. It made his return to the normal hospital situation disappointing and difficult. He writes: 'Now I must return to the 'box system' again. For it seemed to me as if behind the horizon of the cosmos a three-dimensional world had been artificially built up ...'[17] — again that theme of the cosmos without horizon, without partitions, as if a deeper layer of existence, 'the foundation of the world of objects.'

A particular moment in the three-week course of Jung's vision needs remarking upon. He felt the presence of 'inexpressible sanctity' that had a 'magical atmosphere.'[18] 'I understood then why one speaks of the odour of sanctity, of the 'sweet smell' of the Holy Ghost.'[19] Here, I want to use poetic licence by inviting Heidegger to explicate Jung's moment of sanctity, the presence of holiness in the hospital room. Heidegger writes: 'Blue is not an image to indicate the sense of the holy. Blueness itself is the holy, in virtue of its gathering depth which shines forth only as it veils itself.'[20] Robert Avens explains Heidegger's holiness of blue:

> Holiness is not a property of a God ... but a name for all entities insofar as they display a numinous aspect; it is an ingredient that awakens, ensouls, and vivifies everything. Specifically, the holy [in Heidegger] is identified with the blueness of the sky.[21]

The next tale of Jung's encounter with blue occurs in Ravenna on entering the Baptistry of the Orthodox. 'Here, what struck me first was the mild blue light that filled the room ... I did not try to account for its source, and so the wonder of this light without any visible source did not trouble me.' It was here that Jung and his companion envisioned 'four great mosaic frescoes of incredible beauty ... and to this day I can see every detail before my eyes: the blue of the sea, individual chips of the mosaic ...'[22]

These mosaics on the walls of the Baptistry did not actually exist, though they were seen and remembered in detail by both viewers. The light which introduced the vision was blue; the most vivid of the images: 'the blue of the sea.'

The last of these stories of the azure vault I take from Marcel Proust in *Time Regained*, the concluding part of his many-volumed masterpiece, the author as himself and as character reflecting on his lifelong literary effort recounts the rising of his psyche to joy from 'gloomy thoughts' about the 'life of the mind,' which he calls 'unfertile,' 'boring,' 'tedious,' 'useless,' 'sterile' and 'melancholy.'[23]

While crossing a courtyard he stumbles on an uneven paving-stone, and suddenly the oppressive blue mood becomes visual and visionary: '... a profound azure intoxicated my eyes, impressions of coolness, of dazzling light, swirled around me …'[24] Then:

> ...a new vision of azure passed before my eyes, but an azure that this time was pure and saline and swelled into blue and bosomy undulations, and so strong was this impression that the moment to which I was transported seemed to be the present moment …[25]

No sooner is Proust out of this sentence then the *cauda pavonis* appears:

> ...the plumage of an ocean green and blue like the tail of a peacock. And what I found myself enjoying was not merely these colours but a whole instant of my life on whose summit they rested…[26]

Recording his reflections, he concludes with a cogitation about time past and present, 'and I was made to doubt whether I was in the one or the other.'[27] Experiences that so moved him and had given him such felicity were those that joined past and present, 'outside time,'[28] and the doubt about the life of the mind as writer and 'anxiety on the subject of my death had ceased ... since the being which at that moment I had been was an extra-temporal being.'[29] 'The being which had been reborn in me ... with a sudden shudder of happiness ... is nourished only by the essences of things ... A minute freed from the order of time has recreated in us, to feel it, the man freed from the order of time.'[30] The azure vision had brought together the pleasures of the world and the life of the mind, placing time within the timeless, the timeless within time. And with joy he can now say, 'My appetite for life was immense.'[31] Proust presents this azure vision in temporal terms, where time's ineluctable continuity, as in a flash of lightening, is intersected by the joyful certitude of his extra-temporal essence beyond the reach of death. [32]

The harmony of world and mind, resolving doubt and death, brings us to Wolfgang Pauli's dream vision of the world clock, a centerpiece of Jung's 1935 Eranos Lecture published more fully in Jung's Terry Lectures at Yale, *Psychology and Religion,* and later in his *Psychology and Alchemy* (Vol. 12).

I hope the reader's familiarity with this piece of Jung's work allows me to recapitulate only that component of Pauli's vision bearing on our theme: the vertical blue disc that intersects the horizontal one, each disk having its own pulse or time-rhythm.

A letter from Pauli to Jung — in the volume of their correspondence introduced by Beverly Zabriskie — shows Pauli still working on the world-clock dream of several years before. October 15, 1938, Pauli writes: '...I have come to accept the existence of deeper spiritual layers that cannot be adequately defined by the conventional concept of time. The logical consequence of this is that death of the single individual in these layers does not have its usual meaning, for they always go beyond personal life.' Pauli emphasizes the 'sense of harmony' bestowed by the word-clock vision, much as Proust wrote of joy and an appetite for life and Fechner of the beauty of the garden of flowers.

What intersects, breaks into the normal (to use the word from Breuer) world of time's minutes and three-dimensional existence, what moves one outside that 'box' (to use Jung's language of his own vision after the heart attack) is blue's verticality. The break in Proust's step-by-step forward motion, his stumble, parallels Jung's late strange definition of God: 'This is the name by which I designate all things which cross my willful path ... all things which upset my subjective views, plans, intentions and change the course of my life for better or worse.'[33]

Proust's was the last of my stories — last of those I have collected so far. There are surely more waiting in the wings. So now the temptation arises to form the scattered occurrences into a metaphysical conclusion. It would be expected now to leave the earth for the blue yonder and a literalization of the spirit, outside of time, outside the body. Don't the stories support a hypothesis of earthly transcendence? Have we not been encountering the Celestial Kingdom, the 'effulgent blue light of the Buddha body,' visitations of Sophia, of Mary in her blue dress, the ultimate *anima?*[34]

Not quite; not yet; not today. I shall go on insisting until I am blue in the face that Miles' music stayed dark, that blue is the colour of the deep, as Heidegger says, that the poetic AE was earthborn and earthbound, invited by the United States Department of Agriculture to lecture on rural economy; that Cézanne stood day in and day out in the fields among rocks and mountains, painting peasants and apples of this earth; that Novalis, whose degree was in mining, felt called 'to cultivate the earth'[35] and that he took the pen name 'Novalis' from the Latin for 'newly ploughed field;'[36] and that Fechner acknowledged the earth's consciousness to be far superior to that of humans; and that Proust captured every fibre of the earthly emotions in nature, in body, in taste, touch and smell.

If spiritual ascension is not my intention with these tales, what am I intending to convey? First of all, I am elaborating a method for the psychology of *storytelling*. Stories claim neither proof nor truth. Instead of argument, anecdote — individual cases circumambulating a theme. The theme? The *caelum* of alchemy in actual lives, particularly lives open to fresh perception. The method follows Jung's method of *amplification*: building the power of a theme by amplifying its volume with similarities, parallels, analogies. The method is also *empirical* in that it starts and stays mainly with actual experiences. Further, the method is *phenomenological*: let the event speak for itself, bracketing out concepts of spirit, of the numinous, the *coniunctio* and the self.

Most valuable of all, I believe, is the aesthetic of the method which I am attempting. I am employing a rhetorical device, *peitho*, as the Greeks sometimes called Aphrodite, to invite, seduce, charm, enhance and convince by rhetorical, even poetic, means. An aesthetic method relies on texture, images, language, emotion and sudden mysterious irrationalities. The method complies with, and submits to, the content. Story keeps its words close to events, logos in the embrace of psyche. Like blue itself, an aesthetic method conceals and reveals, withdraws from our prehension, tempts us to follow after it, and connects invisibly, analogously, all the stories and persons existing together in the same field. The method is relieved of interpretations and personalistic contexts. It aims to present things as they are and also as played upon the blue guitar – to use Wallace Stevens's famous phrase,[37] following also his statement: 'I am thinking of aesthetics as the equivalent of *apercus* which seems to have been the original meaning.'[38] Sudden openings of the heart and mind and senses, especially of the eyes; insights, ahas, analogies, unique epiphanies that shake the soul, carry it to the edge, and free it from the box.

The box is also psychology: not psyche, but the '-ology,' that parasitical suffix that sucks the psyche dry. Long before there was psychology there were tales, old wives' tales, grandmothers' tales, oral accounts of origins and great deeds, theatre of tragedy and comedy, the gossip of the day carried by messenger, lessons learned at the feet of a teacher, stories passed down rich in the ways of the world and the ways of the soul. Long before psychology there were the bedside observation of physicians, of captains on the field of battle, painters of portraits, breeders of animals and trappers, of midwives and judges and executioners. Psychology's case reports are too often botched attempts to continue the storytelling tradition. Too soon we draw theoretical conclusions obliged by '-ology' to package psyche in a box. We would win from every story the trophy of meaning.

Ideally, an aesthetic method, if I may call it that, would let the beauty of an event, its sweet shock, instruct the soul, educate it by leading it to an edge, out of the box of the already conceived, and into pondering and wondering. The method suits the correspondences that compose the cosmos itself, each thing implicating other things by likeness rather than by causality, in an implicit order of the world. Metaphors and analogies abound. The display of images addresses the 'poetic basis of mind' which is our most native mode of comprehension.[39]

Alchemy caught me and taught me with its aesthetics — its colours and minerals, its paraphernalia, freaks and enigmatic instructions. It is like a vast collective artwork built through centuries. It offers an aesthetic psychology: a myriad of apercus, images, sayings, stories, formulae; and all the while engaged with the matters of nature. It tells us to throw away the book of conceptual systems; no need for male and female, typology, stages, opposites, transference, self. Conceptual systems may be useful as scaffolding for better access to the *massa confuse* which alchemy presents to a logocentric mind. Too soon, however, the conceptual scaffold replaces alchemy itself reducing it to merely providing examples to support the conceptual scaffold. *Que lastima!*

Allow me one more story from Jung. This is the moment in the articulation of our field when Anna O's poetic blue is saved from Breuer's prosaic brown.

It is the moment of Jung's 'steep descent,' when he felt he was 'in the land of the dead.' 'The atmosphere was that of the other world.' First, he met Salome and Elijah, which he interpreted as Logos and Eros, but then retreated from this intellectualization. Then, 'another figure rose out of the unconscious ... I called him Philemon.'[40] Jung had already met Philemon in a dream:

> There was a blue sky, like the sea, covered not by clouds but by flat brown clods of earth. It looked as if the clods were breaking apart and the blue water of the sea were [*sic*]

> becoming visible between them. But the water was the blue sky. Suddenly there appeared from the right a winged being sailing across the sky ... He had the wings of the kingfisher with its characteristic colours.[41]

If you know these pages in *Memories, Dreams, Reflections,* you will recall that it was Philemon especially, Jung said, who taught him that 'there are things in the psyche ... which produce themselves and have their own life.' 'It was he who taught me psychic objectivity, the reality of the psyche.'[42] Jung later placed a kingfisher's wing in one of his paintings, a wing whose colour has given its name to a particularly brilliant blue. That blue with its shimmer of coppery gold recalls Stevens' poetic way of stating psychic objectivity: 'When the sky is so blue, things sing themselves.'[43] The blue wing sailing across the sky announces in an aesthetic image the arrival of Jung's new knowledge of the autonomy of the psyche.

The autonomy of the psyche is preserved by an aesthetic method. Jung's term 'the objective psyche' refers to more than the spontaneous production of 'internal' events. Psychic events are not atomistic particulars only; they bode forth analogous affinities such as Freud sought in free associations. Events 'sing themselves' further, dream the myth onward, even 'infinitely' as Mueller's poem says, and in this way they are objective, freed from the given — the psychic feeling or fact or fantasy — by their analogous implications. Events dissolve their own edges and overreach themselves, creatively objectifying psyche in the production of complex forms that bring their own norms which seem to usual judgments as irrational, amoral or abnormal, giving rise to the prejudice that the aesthetic and the ethical are incommensurables.

To the great misfortune of our tradition, Josef Breuer could not hear the transposition to psychic objectivity, the singing of things themselves. Despite his arduous devotion to his case he did not recognize the poetic basis of mind. He called Anna O's seeing blue a 'secondary state' invading the 'more normal view.' Her dress was simply brown. Yet, Anna's fantasies were breaking into, and her out of, the historical, literal and personal. Her symptoms were mainly bodily and the analysis was wrapping her body in blue material. Breuer describes Anna as 'markedly intelligent with penetrating intuition [and] powerful intellect ... She had great poetic and imaginative gifts.'[44] The patient was being led by her very eyes and the call of her symptoms to follow the blue in keeping with her gifts.

The blue incursion early in the case of Anna O, upon whose psyche our field is founded, requires us to make a small correction to the poem about Monet's cataracts with which I began. Lisel Mueller has Monet saying: 'I tell you it has taken me all my life to arrive at the vision of gas lamps as angels. Fifty-four years before I could see Rouen cathedral is built of parallel shafts of sun....' Fifty-four years! Jung, too, places the *caelum* at the end of his *opus maior* and the end of his scholarly life. And an implication from Proust requires a correction. Is it only as death enters our thoughts and we are near the last page of that novel's *via longissima* that the azure vision, with peacock tail, flowers, and joy finds us?

Others tell a different story. Miles Davis as a tiny boy; Little AE with the silver and blue of the magic sword; Anna O barely out of her teens. Fechner in his early forties; Jung with Philemon also in mid-life — not at the end of an *opus contra naturam*, a work that struggles to arrive. Rather, we learn that the 'arrival,' if it be that, is outside of time altogether and what is outside of time cannot be achieved through time. Either the achievement is never reached — analysis 'interminable' to use Freud's term — or, timeless, is eternally present from the beginning and potential in each moment of the work. Far out is where we start from! Hence, *caelum* is one of the names of the *materia prima,* the starting stuff and permanent basis of the work.[45] The striving toward healing on the *via longissima* is present from day one in the very fantasy of wholeness as an impulsion of the *caelum* as *unus mundus.* The fantasy that things are improving, integrating, and Jung's synthetic or prospective method too, are modes of stating in terms of time the *caelum* that is always there as the *mundus archetypus* to which all things desire to return to their potential.

Practitioners of analysis who carry an alchemical imagination in their devotion will keep faith with the primordial firmness of their vision. Like Fechner they will wear blue glasses and like Cézanne they will hold a blue brush in their hands, hearing things as they are played upon Stevens' blue guitar. Practitioners will be remembering that practice itself is an aesthetic activity that awakens the soul from anesthesia by revisioning from the first hour. And, like Miles Davis, they will feel the blue flame pushing forward to some frontier, the edge, and out of the box, released from the logic of opposites and the coercion of centering, without tops and bottoms and lines of horizons. Certitude and Trepidation, both; and like Proust, with an immense appetite for life.

All along, our blue theme has been haunted by nostalgia, *Heimweh*, a nest from which we have flown, the harbour where we have not arrived, a longing that imagines an elsewhere that is 'not here.' Not here: that is the essential plaint of the émigré, of the mercenary by the campfire in a foreign land, of the displaced, the exiled. A mood of wishing, 'mixing memory and desire,' and the disconsolate regret of heartbreak. Nostalgia gives a particular hollow ache to the 'missing blue' mentioned by Jung in his *Psychology and Alchemy*, and the missing blue in the *opus* itself, dominated by black, white, yellow and red.[46]

Blue withdraws from us, says Goethe; it is the absence shadowing the alchemical process, the essentially missing. Not-hereness profoundly motivates the work all along the way: not enough, not right, not fulfilling, something else, something more. Intensely present in its utter absence, reminding the soul of its exile. The inevitability of exile as the necessary ground for removal of all supportive identities: the very idea of identity, of self-identity, of self itself are the structures and straws to which loneliness clutches. Exile reveals that we are each foundlings and that there is no other home but the cosmos itself from which no single particle can be severed, and to which each and every thing belongs and homecoming continually takes place, occurring in our very breathing its blue air. Cézanne drenched all things in blue, keeping them from separateness, making visible 'how heaven pulls earth into its arms.' The azure vault folds hard edges into its cosmic comprehension. No elsewhere, no exile, no nostalgia.

Precisely here we can make a distinction, on the one hand, between the blue of the *unio mentalis* that occurs, writes Jung, before the *unus mundus* and, on the other, the celestial blue of the *caelum*. That first blue is more mood than effulgence. Its blue roses are intertwined with lunar subjectivity, called 'pleurosis' in the case of Laura in Tennessee Williams's *Glass Menagerie*. That first blue emerges as the *nigredo* clears into the *albedo* and the mute mind finds voice, lightens up and can sing the blues, express the melancholy. That blue I explored at length in an essay twenty-five years ago.[47] This azure blue is the 'visible spirit of Song,' wrote Novalis, the source of song itself according to AE, the blue of vision further than reflection.[48] If the *unio mentalis* signifies the confluence of understanding and imagination, an understanding by means of images, the *caelum* is beyond understanding — though we may work at it as did Pauli. It feels unimaginable, incomprehensible — as to Jung after Ravenna. Magical. It simply happens, out of the blue, simple and evident and truthful as the sky happens, as death happens, unfathomable and undeniable both. A universal given, a gift. The eyes no longer able to grasp what they see; the eyes become the ungraspable air by which they see. It is vision.

Science fosters the separations, the exile. Science, its root, *scire* (to know), has a further root in Greek, *schizo* (cleft, splinter, separation) and further, Sanscrit *chyati* (divides). Instead of science, why not *séance* for these sessions that invoke our common ancestor? Séance is defined by the dictionary as a meeting of a learned society and also a meeting that attempts to connect with the dead. Jung's expansive vision in hospital took him to the edge of death. Return to life meant divisions, separations: 'the grey world with its boxes.'[49] But there are other ways out of the box, other ways for the grey world to discover a blue vision.

In this vision the world appears as analogies. All things refer, imply, connote. Likenesses everywhere and so things cure one another by means of similarities. 'Objects existing together' (Badt on Cézanne). Edges banished, says Monet in Mueller's poem: 'I will not return to a universe of objects that don't know each other.' 'The human soul recognizes itself in the world, as the world.'[50] Descartes' separations dissolve. Analogy, says William James, is the mark of Fechner's method and genius, for it reaches in all directions and finds subtle strands of implications.[51] 'Attentiveness to subtle signs and traits …' writes Novalis.[52] The method of discovering analogies carries further than symbols, further than images as such. It is a poetic connection. Reich might add, 'an erotic connection.' 'Poetry,' writes Wallace Stevens, 'is almost incredibly one of the effects of analogy.'[53] So, incredibly, is healing. Analogy disrespects definitions, leaps over defences, listens through walls to overhear meanings. The poetic mind resolves the need for meaning, Jung's underlying reason for psychotherapy. 'The poet is the transcendental doctor,' wrote Novalis.[54]

James goes on to describe Fechner's *unus mundus*: 'All things on which we externally depend for life — air, water, plant and animal food, fellow men, etc. — are included in her [the earth] ... She is self-sufficing in a million respects in which we are not so.'[55] Then, like an astronaut's vision of the globe, and like Jung's vision from his hospital bed, James captures Fechner's vision with this paragraph: 'Think of her beauty — a shining ball, sky-blue, sunlit over one half, the other bathed in starry night ... she would be a spectacle of rainbow glory ... Every quality of landscape that has a name would be visible in her at once ... a landscape that is her face' — Novalis saw the face of the blue flower. 'Yes,' writes James, 'the earth is our great common guardian angel who watches over all our interests combined.'[56]

Just this Fechner perceived through his blue glasses, the *unus mundus*, the earth as angel, the deep ecology of the Gaia Hypothesis as truth because seen and felt, not because believed in or scientifically buttressed. 'We are called to form the Earth,' wrote Novalis.[57] 'Doctor, if you could only see.' This vision is the *caelum* experience, and without vision, continues James famously at the end of his rapture recapitulating Fechner's rapture, 'without vision the people perish.'[58] And we perish, the patients perish, our psychology perishes without reminiscence of the vision that impelled us to Jung in the first place, a vision that is there from start to finish as *prima materia* and *unus mundus*, a recollection that gives reason for Jung's turn to alchemy for the amplification and substantiation of his life's extraordinary work, leaving us with the charge to recollect that our work, however boxed and clocked, however bandaged our eyes, is always under an azure vault.

Notes

1 Ruland, *A Lexicon of Alchemy,* p. 10.
2 Jung, *Collected Works,* Vol. 14, pp. 761–770.
3 Breuer, 'Case Histories: Fraulein Anna O,' in Sigmund Freud, *The Standard Edition of the Complete Psychological Works,* II, p. 33.
4 Lowrie, *Religion of a Scientist: Selections from Gustav Th. Fechner*, p. 211.
5 Ellenberger, *The Discovery of the Unconscious,* pp. 217, 512.
6 AE, (George William Russell), *The Song and its Fountains,* pp. 12–13.
7 Davis, Miles (with Quincy Troupe), *Miles: The Autobiography*, p. 1.
8 Bamford, *An Endless Trace,* p. 228.
9 Bamford, *An Endless Trace,* p. 229.
10 Bamford, *An Endless Trace,* p. 229.
11 From *Heinrich von Ofterdingen* (by Novalis), 'Eros and Fable: birth of Astralis,' in Christopher Bamford, *An Endless Trace,* p. 230.
12 Adams, *The Fantasy Principle,* p. 89.
13 Badt, *The Art of Cézanne,* p. 82.
14 Badt, *The Art of Cézanne,* p. 57.
15 Badt, *The Art of Cézanne,*p. 56; cf. Zola, *L'Oeuvre.*
16 Jung, *Memories, Dreams, Reflections,* p. 270
17 Jung, *Memories, Dreams, Reflections*, p. 273
18 Jung, *Memories, Dreams, Reflections*, p. 275.
19 Jung, *Memories, Dreams, Reflections*, p. 275.
20 Heidegger, *On the Way to Language*, p. 166.
21 Avens, *The New Gnosis: Heidegger, Hillman, and Angels,* p. 56.
22 Jung, *Memories, Dreams, Reflections*, pp. 265–66.
23 Proust, 'Time Regained,' in *Remembrance of Things Past*, Vol. III, part 3, p. 898.

24 Proust, 'Time Regained,' p. 899.
25 Proust, 'Time Regained,' p. 901.
26 Proust, 'Time Regained,' p. 901.
27 Proust, 'Time Regained,' p. 904.
28 Proust, 'Time Regained,' p. 904.
29 Proust, 'Time Regained,' p. 904.
30 Proust, 'Time Regained,' pp. 905–6.
31 Proust, 'Time Regained,' p. 905.
32 Proust, 'Time Regained,' p. 905.
33 Jung, *Letters*, Vol. 2, 5 December 1959.
34 Jung, *Collected Works,* Vol. 11, p. 852.
35 Bamford, *An Endless Trace,* p. 198.
36 Bamford, *An Endless Trace, idem.*
37 Stevens, *The Collected Poems*, p. 165.
38 Stevens, *Letters*, p. 469.
39 Hillman, *Revisioning Psychology*, p. xvii.
40 Jung, *Memories, Dreams, Reflections*, p. 175.
41 Jung, *Memories, Dreams, Reflections*, p. 176.
42 Jung, *Memories, Dreams, Reflections*, p. 176.
43 Stevens, 'Debris of Life and Mind,' *The Collected Poems*, p. 338.
44 Breuer, 'Case Histories: Fraulein Anna O,' in Sigmund Freud, *The Standard Edition of the Complete Psychological Works,* II, p. 21.
45 Jung, 'Lecture Notes,' 20 June 1941.
46 Jung, *Psychology and Alchemy, CW* 12.
47 Hillman, 'Alchemical Blue and the Unio Mentalis.'
48 From *Heinrich von Ofterdingen* (by Novalis), 'Eros and Fable: birth of Astralis,' in Christopher Bamford, *An Endless Trace,* p. 229.
49 Jung, *Memories, Dreams, Reflections*, p. 275.
50 Bamford, *An Endless Trace,* p. 230, on Novalis.
51 James, 'Concerning Fechner,' in *A Pluralistic Universe,* p. 151.
52 Bamford, *An Endless Trace,* p. 224.
53 Stevens, 1951, p. 117.
54 Bamford, *An Endless Trace,* p. 220.
55 James, 'Concerning Fechner,' in *A Pluralistic Universe,* p. 157.
56 James, 'Concerning Fechner,' in *A Pluralistic Universe,* p. 164.
57 Bamford, *An Endless Trace,* p. 220.
58 James, 'Concerning Fechner,' p. 165.

Bibliography

Adams, Michael Vannoy, *The Fantasy Principle* (Hove & New York: Brunner-Routledge, 2004).

AE (George William Russell), *The Song and its Fountains* (Burdett, New York: Larsen Publications, 1991).

Avens, Roberts, *The New Gnosis: Heidegger, Hillman, and Angels* (Putnam, CT: Spring Publications, 2003).

Badt, Kurt, *The Art of Cézanne*, Sheila Ann Ogilvie, trans. (New York: Hacker Art Books, 1985).

Bamford, Christopher, *An Endless Trace* (New Paltz, NY: Codhill Press, 2003).

Breuer, Josef, 'Case Histories: Fraulein Anna O.' in Sigmund Freud, *The Standard Edition of the Complete Psychological Works*, II (London: Hogarth, 1955).

Davis, Miles (with Quincy Troupe), *Miles: The Autobiography* (New York: Da Capo, 1989).

Ellenberger, Hénri F., *The Discovery of the Unconscious* (London: Allen Lane, The Penguin Press, 1970).

v. Goethe, J. W., *Goethe's Color Theory*, Rupprecht Matthei and Herb Aach, eds. (New York: van Nostrand Reinhold, 1971), p. 170 (p. 310 of the Eastlake translation).

Heidegger, Martin, *On the Way to Language*, Peter D. Hertz, trans. (New York: Harper & Row, 1971), p. 166, cf. 169–70. [German: *Unterwegs zur Sprache*, (Pfullingen: Neske, 1965)].

Hillman, James, *Revisioning Psychology* (New York: Harper & Row, 1975), p. xi.

Hillman, James, 'Alchemical Blue and the Unio Mentalis,' *Sulfur*, 1 (1981), pp. 33–50.

James, William, 'Concerning Fechner,' in *A Pluralistic Universe* (London: Longmans, Green, 1909).

Jung, C. G., *Collected Works* (Princeton: Princeton Univ. Press, and London: Routledge & Kegan Paul).

Jung, C. G. (with Aniela Jaffe), *Memories, Dreams, Reflections* (London: Collins/ Routledge & Kegan Paul, 1963).

Jung, C. G., 'Lecture VIII, 20 June 1941,' in *Alchemy: E.T.H. Lecture Notes*, Barbara Hannah, ed. (Zürich, 1960).

Jung, C. G., *Letters*, Vol. 2 (Princeton: Princeton Univ. Press, 1975).

Kandinsky, Wassily, *Concerning the Spiritual in Art*, M. T. H. Sadler, trans. (New York: Dover, 1977), p. 38.

Lowrie, Walter, *Religion of a Scientist: Selections from Gustav Th. Fechner* (New York: Pantheon, 1946).

Meier, C. A., (ed.), *Atom and Archetype: Pauli/Jung Letters*, David Roscoe, trans., (Princeton: Princeton Univ. Press, 2001).

Mueller, Lisel, *Alive Together* (Baton Rouge, LA: Louisiana State Univ. Press, 1996).

Proust, Marcel, 'Time Regained,' in *Remembrance of Things Past*, Vol. III, part 3, Terence Kilmartin and Andreas Mayor, trans. (New York: Vintage Random House, 1982).

Ruland, Martin, *A Lexicon of Alchemy,* A. E. Waite, trans. (London: Watkins, 1964).

Stevens, Wallace, *Letters*, Holly Stevens, ed. (New York: Alfred A. Knopf, 1972).

Stevens, Wallace, *The Collected Poems* (New York: Alfred A. Knopf, 1978).

Part II

Sky and Psyche

3. Chartres Cathedral and the Role of the Sun in the Cathedral's Christian Platonist Theology

BERNADETTE BRADY

Chartres Cathedral is a treasury of esoteric and spiritual western knowledge and, given the scale of its encyclopaedic structure, this paper can only focus on a small fraction of the religious and sacred significances contained within its stone, glass and orientation. This paper argues that the combination of Gothic architectural dimensions and the nature of twelfth-century stained glass created a sacred place in Chartres Cathedral, which was not only in keeping with the twelfth century themes of Christian Platonism but was — and still is — evocative of the actual experience of Platonistic mysticism. Furthermore, this paper argues that the unusual orientation of the cathedral and the annual movement of the sun may well have been incorporated in the design of the current cathedral to enhance this theology.

> I said to my soul, be still, and let the dark come upon you
> Which shall be the darkness of God …
>
> T. S. Eliot, 'East Coker'[1]

A brief history of Chartres Cathedral

The cathedral stands at the top of a mound which overlooks the village of Chartres in northern France. This mound has been considered sacred from pre-Roman times, and Charpentier argues that it was the site of a dolmen sacred to the Druids.[2] Upon this sacred mound the earliest recorded church, a Gallo-Roman chapel, was built between the fourth and sixth centuries. This chapel was thought to be modelled on the Church of the Holy Sepulchre and built in the fourth century, supposedly over the tomb of Jesus Christ in Jerusalem.[3] There have been five known

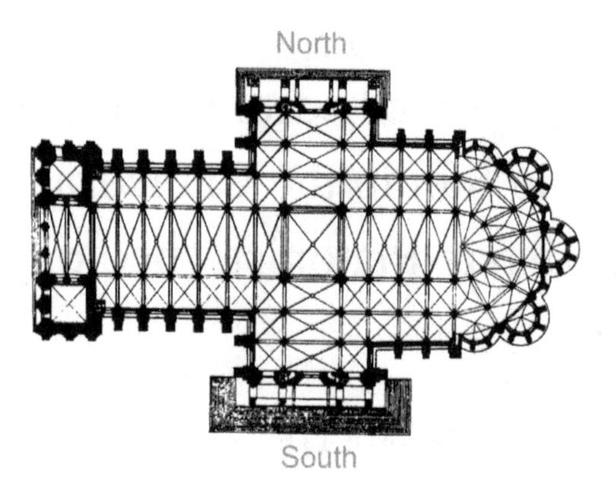

Figure 1: Ground plan of Chartres Cathedral. The labels of South and North are not true compass points but simply the traditional manner by which the two sides are named.

churches built on the site of Chartres Cathedral. All of these have been destroyed by fire except the last. In 858, when the church had been burnt and pillaged by Vikings, Charlemagne's grandson, Charles the Bald, presented the newly rebuilt church with a sacred relic known as the *Sancta Camisia,* which was said to be a garment worn by Mary when she gave birth to Christ.[4]

The current Chartres Cathedral was built between 1194 and 1220 and is the work of just one generation. It was constructed in the fever of French public works, which included cathedrals, churches, abbeys, bridges and town walls, and which consumed France for forty years from the 1190s to the 1230s.[5] Its grandeur and scale was made possible by donations of money and labour from across France, an outpouring of gifts which was fuelled by the 'miracle' of the survival of the relic of the *Sancta Camisia* from the fire of 1194. The survival of the relic was also taken as a sign that the site was indeed the location of the Virgin Mary's earthly throne.[6]

The current cathedral was constructed keeping the earlier Romanesque church's Royal Portal (the western doors) and this, along with the location of the crypt in the east, established its orientation as well as length (see Figure 1). The alignment of Chartres Cathedral is, therefore, far older than its construction date of 1194. Indeed the ground plans of all five churches or cathedrals built on the site fit inside, or on, each other like Russian dolls, with each on the same alignment as the previous.[7]

Christian Platonism and architecture

The eight hundred years of the cathedral's existence have bought changes in Christian religious philosophy. In twelfth-century France, as in most of Europe of that time, Platonic orientation towards God suggested that he was hidden and revealed Himself through the application of one's mind.

In a letter to Dorotheus, the Christian Platonist, Dionysius, wrote:

> The Divine gloom is the unapproachable light in which God is said to dwell ... He is after all the object of sensible and intelligent perception, and saying in the words of the Prophet, [St Paul] 'Thy knowledge was regarded as wonderful by me; It was confirmed; I can by no means attain unto it.[8]

In the second century, Clement of Alexandria is credited with firmly embedding neo-Platonism into Christian theology. His work was devoted to continuing this theme of finding or building a relationship with God through the use of one's mind, *ascensio mentis in Deu* — the ascent of the mind to God.[9] St Augustine, in the fourth century CE, linked Christianity's specific belief in incarnation and one's consequent salvation to Platonism. Augustine took the philosophical reflections of Platonism and made them practical in terms of Christianity.[10]

When Chartres Cathedral was completed, Christian theology was still strongly neo-Platonic. Education and the illumination of the truth were established as the road to knowledge of God, and ignorance was believed to be a form of evil. God dwelled in the darkness and through seeking the unknowable one could find God.

By the eleventh century, Christian centres like Chartres were devoted to becoming places of education, focusing on the seven Liberal Arts which were: the *Quadrivium* (arithmetic, music, geometry and astrology) and the *Trivium* (grammar, rhetoric and logic). Indeed, Chartres, under the leadership of Bishop Fulbert, the bishop at the time of the construction of the Royal Portal in 1020, became one of the great centres of learning and was famous across Europe.[11]

The Gothic style and its role in promoting Christian Platonism

The Gothic style was originally called the New Style and was promoted by Abbott Suger of the Abbey of St Denis, who believed the Gothic arch allowed for a greater mystical experience by the beholder. His words concerning the New Style hover between the poetic and the devotional:

> ... then it seems to me that I see myself dwelling, as it were, in some strange region of the universe which neither exists entirely in the slime of the earth nor entirely in the purity of heaven; and that, by the grace of God, I can be transported from this inferior to that higher world in an anagogical manner.[12]

The more rounded Romanesque arch was simply thought to move energy from the earth on one side of the arch, back into the earth on the other side of the arch. In the twelfth century this configuration represented death; earth to earth, dust to dust. The one-fifth Islamic arch adopted by Gothic architects, however, contained a point at its top and hence, according to Strachan, was thought to move the energy from the earth on both sides of the arch to the heavens through its peak, thus representing life.[13]

Geoffrey de Leves, the twelfth-century bishop of Chartres, became inspired by Abbott Suger's New Style architecture in the Abbey of St Denis and, after the fire of 1194, decided to rebuild Chartres in this new form.[14] At that time it was considered that the Holy Ghost, mysterious and unreachable, was the Platonic world-soul and de Leves wanted this view to be captured in the very architecture of Chartres itself. The new cathedral was, therefore, to express in its very form the views of Christian Platonism, and to evoke in the visitor the experience of a hidden God who is dwelling in the sacred darkness, a God who can be discovered through the pursuit of knowledge.[15]

The New Style lent itself to mystery and awe; it was space created with a compass rather than a ruler. While the architecture of earlier Romanesque churches was based on arithmetic and on whole numbers, the New Style was based on geometric ratios, such as one to the square root of three — ratios that cannot be constructed or understood with a ruler and a slide rule.[16] The study of the proportions of this New Style consequently lead to an encounter with mystical number ratios, which beckoned towards deeper levels of learning and ended in such mythic places as the golden mean. Strachan sums this up elegantly:

> In moving from measurable arithmetic to immeasurable geometry, they reflected a new desire to express the ultimate incomprehensibility of God … It also constitutes the difference between an architectural style [Romanesque], which housed a spirituality in which the mystical was present but marginal and subservient to the liturgy of the mass, and one in which it was central to the *raison d'être* of enhancing the possibility of mystical experience in the building itself, with or without the liturgy of the mass.[17]

This is architecture reflecting a form of mystical theology, the immeasurability of God, the mystery of space where God is hidden from consciousness, where God needs to be revealed. Indeed, God Himself is considered to be the architect of Chartres Cathedral, the *elegans architectus* who built the cathedral to reflect His cosmos.[18] This sacredness and divinity, inspired by the cathedral's form and shape, is captured by Strachan in his ponderings upon the interior of Chartres:

> … that Chartres is still a sacred space which, without any religious service taking place, can deeply affect those who visit, so that it is possible to say that the building itself can produce a spiritual experience; and secondly, that the spiritual power of the place is somehow connected to the fact that it is dark.[19]

Indeed, von Simson states that the primary motivation of building in the New Style was neither architectural innovation nor aesthetic beauty, but rather a desire to house a type of mystical spirituality and while stone can beget form which inspires awe and wonder, the real drama of Christian Platonism is displayed on the lighting stage created by the cathedral's stained glass.[20] As von Simson pointed out, 'According to the Platonizing metaphysics of the Middle Ages, light is the most noble of natural phenomena, the least material, the closest approximation to pure form.'[21]

The nature of the twelfth-century stained glass in Chartres Cathedral

As the mystical number ratios of geometry replaced the logic of arithmetic in the shift from Romanesque to Gothic religious architecture, so stained glass replaced the wall frescoes of the earlier period as a dominant element.

Chartres Cathedral contains 166 windows which make up just under an acre of stained glass.[22] This glass represents an encyclopaedia of twelfth and thirteenth-century Christian theology and is largely intact. The damage that has occurred to the glass happened in the eighteenth century in an effort to bring more light into the cathedral. In 1753, in an attempt to light the choir, the borders of several thirteenth-century windows were removed and replaced with clear glass. In 1773, eight of the choir windows and four in the transept were smashed, also in attempt to bring more light into the place where the bishops stood.[23] Other Gothic cathedrals have tragically lost large parts of their glass, sometimes all of it, with the result that more recent windows cast their Renaissance, or at times modern, lighting plans across the original Gothic stage. Such clashing of lighting plans effectively obliterates the older, more delicate, lighting designs of the twelfth-century creators of the Gothic cathedral.[24]

Stained glass by its nature does not benefit from surface light — light shining directly upon its face — which makes the glass appear flat and dull.[25] Stained glass needs lighting from behind so that it can transmit the light, or be illuminated by it, allowing maximum visual impact of the glass. In Chartres Cathedral the internal illumination was, and still is, potentially quite low. Normal sunlight at midday has a candle power of 8,000 to 10,000 footcandles (the measurement of light level equivalent to the light intensity made by one candle at a distance of one foot); the average illumination levels inside Chartres Cathedral are one to two footcandles — the brightest readings being closer to five footcandles, while other readings in the north ambulatory are well below one footcandle.[26] Taking into account the fact that weathering over the last eight hundred years would have reduced the ability of the glass to allow light into the cathedral, this has been balanced by the removal of twelve windows and the installation of various clear panels to allow in more light, so one can probably assume that the levels of illumination in the cathedral today are similar to what was intended in the twelfth century.

Such low lighting levels alter the effectiveness of the human eye. Johnson has pointed out that when the eye adapts to low illumination, known as scotopic vision, it moves into a state of relative myopia where it relaxes and tends to have a fixed focus.[27] The ability of the human eye to focus on a given colour at low light levels is defined as the colour's acuity rating. The colours yellow, yellow-green and green have the highest acuity ratings which means that in low light conditions the human eye can still focus on these colours. At the lower end of the acuity scale are the colours of blue-greens, deep reds and blues, with the blues having the lowest acuity rating of all. This means that at low light levels the human eye finds it difficult to adjust its focal length to these colours, and these low end acuity rating colours are in fact the dominant colours of the twelfth and thirteenth-century glass in Chartres Cathedral.

As a result of the use of these colours in the cathedral's glass, combined with the intentionally dim lighting environment, our eyes enter into a mildly myopic state and are unable to focus clearly on the coloured glass. The eye continues to seek the correct focal length by shifting it back and forth in the dim surroundings. As a result the images appear to float off the plane of the glass, seemingly stepping off the window and hovering in front of the observer's eyes. This is a conscious twelfth-century trick of colour and lighting which has had a profound mystical impact on tens of thousands of visitors to the cathedral over the last eight hundred years.[28]

Other authors, unaware of the designers' intent in terms of lighting and imagery, become overwhelmed with the mystical impact. Even James, whose work is focused on the stone and the structure of the cathedral, comments on the visual impact of the glass: 'From inside we can still glimpse this today in the fractured prismatic colours of the stained glass that transforms that solidity into a skimmer that floats rather than sits.'[29]

Johnson suggests that this effect was intentional and that the glassmakers of Chartres purposely designed the glass to transmute rather than transmit rays of light.[30] Glass, which transmutes light, is achieved by varying the glass density as well as using foreign matter in the glass. This transmuting property allows the windows to glow, unlike transmitting glass that would allow shafts of light into the interior. Medieval glassmakers were capable of creating clear glass, but the intent of the glassmakers with the Cathedral's glass was to enhance the diffuseness. Indeed, Johnson likens the Chartres glassmakers to medieval jewellers who wanted the windows to act like a gem and to glow with light rather than simply allow it to enter the cathedral.[31] Thus, the unique lighting plan of Chartres Cathedral was made up of a combination of dim lighting, the use of blues and reds, and glass which transmuted rather than transmitted light. This combination produces visual disorientation to the human eye and thus a window contemplated by a visitor would, as their eyes endeavoured to find the correct focal length, give out a celestial glow and allow the images to appear to move off the glass in an almost holographic manner.

In contrast to this delicate twelfth-century lighting plan, stained glass with a predominance of yellow and green colours is far more comfortable to the human eye, which can readily hold its focus, giving greater visual certainty. These are the major colours in the stained glass of the Renaissance buildings and as Johnson states:

> The Renaissance, for example, used lighter windows for more clearly illuminated interiors, enhancing three-dimensional form and revealing the classic order, the carefully adjusted proportions, and the clarity of the building's scale. If man was to be in the centre of the world, he wanted to be seen there, in the midst of his works.[32]

However, following recent additions to the cathedral's lighting scheme, the interior is now illuminated with numerous spotlights focused on the sculptures and which shine into the eyes of those who seek to observe the glass. By illuminating the sculptures, the current guardians of the cathedral have removed the fragile relationship between the stained glass and the human eye and thereby eliminated the potential for its particular, and unique, experience. One would like to think that in the future the guardians of the cathedral will adopt a greater understanding of the relationship between the glass and light and introduce a more sympathetic lighting scheme, at least periodically.

Chartres Cathedral also displays one of the great colours of medieval art, that known as celestial blue. It only occurs in twelfth-century glass and then, according to Johnson, only rarely. In the interior of the cathedral, celestial blue is the colour with the greatest potential for causing visual confusion for the observer. Johnson describes the colour thus:

> This celestial blue is one of the supreme moments in medieval art, for it is seemingly without body or substance, composed entirely of light. Weightless, ethereal, it is matter

> transformed. ... seen from a distance it seems that this divine colour permeates the entire ensemble, emanating, as it were from the Supreme Light to the lesser lights, transforming all things.[33]

Thus, when we consider the glass of Chartres Cathedral, we must recognize that we are stepping onto a stage where the proportions of stone, sculpture and lighting are specifically designed to propel us on a mystical journey of finding the divine in the sacred darkness. This mystical journey is one saturated with celestial light where divine figures float before one's eyes. It filled early and even modern pilgrims with a sense of a heavenly Jerusalem and potentially could continue to fill twenty-first century tourists with a sense of awe and wonder. With light playing such an important part in the mysticism of the cathedral, it is feasible that the role of the movement of the sun could also have been incorporated into the Cathedral's construction.

The design and orientation of the Cathedral

Chartres Cathedral's visual ratios and proportions emphasize the sense of sacredness and have given rise to a seemingly endless analysis by different writers, all of whom support a variety of different theories of lost or divine knowledge. This is a process that undoubtedly feeds back into the concept of the sacredness of the space which, in turn, produces yet more motivation to intellectually 'know' the mysteries of Chartres. This is, of course, the aim of Christian Platonism: God is hidden and is revealed through knowledge. This seeking of knowledge is stimulated by the New Style of architecture whose detailed examination of proportions leads to never ending unsolvable mysteries.

It can be argued that the mysteries of the numbers and ratios of Chartres are not designed to be solved. Just as, in keeping with Christian Platonism in which God is knowledge which must be explored but can never be *fully* revealed, so the cathedral is designed to draw, pull and attract researchers to probe ever deeper into the magic of space and numbers, to confront the cathedral's deeper design and to bring all research to the same place — a place of mystery, reverence and awe. The builders of the cathedral even appear to warn visitors of their intent with a labyrinth boldly placed in the centre of the western part of the floor. This labyrinth offers only one pathway with all journeys ending in its centre; thus, the labyrinth can be seen as a physical metaphor of the concept that all seeking of knowledge leads to only one place, the one true God.

But there are some more simple and obvious features of the cathedral's design that can be easily grasped. On the northern side, which never receives direct sunlight, are the images and sculptures of the Old Testament, the things of the past, perhaps the things that no longer need to receive the full celestial glow of the sun. This side depicts the history of the church, its roots, its early writings. In contrast, the New Testament and new covenant is embodied on the southern side, which always receives the sun's light.[34] Although this is a simple design feature, it does reflect an obvious consideration of the sun's movement during the course of a year — but this may be just one feature of the sun's role in the design of the cathedral.

The sun and the Cathedral's orientation

The orientation of Chartres Cathedral is 47° north-east which is quite different from the normal east-west orientation of churches across Christendom.[35] The sun in the northern hemisphere (notwithstanding equatorial latitudes) will culminate due south on any given day of the year. The altitude of the sun's culmination is dependant upon the latitude of the location and the actual calendar date. For a given location on a particular calendar date, the sun will always (barring the tiny effects of mutation) culminate due south at the same altitude from one year to the next. Additionally, for a given latitude, the exact location where the sun will rise or set on the horizon will be a constant for a given calendar date. Furthermore, if we watch all the sunrises for a full year, they will always fall in the same band on the eastern horizon, and similarly for sunsets in the west. This band can be measured using a compass and such a measurement is known as the azimuth of the sun's (or any other celestial body's) rising or setting position on the horizon. For the latitude of Chartres, the sun will rise within a band of the horizon which measures from 52° north-east to 52° south-east and set in corresponding locations in the west, as shown in Figure 2. For any given calendar date the exact position of sunrise and sunset can be expressed in degrees of azimuth (traditionally measured in degrees from due north in an easterly direction) and, as already stated, it will be constant from one year to the next.

The solstice axes

Recognizing that the orientation of Chartres Cathedral predates the current structure and follows the same alignment as the earlier sacred buildings on this site, if we consider this orientation with regard to the rising and setting of the sun on key dates of the year, some interesting patterns do emerge.

The solstice axes (see Figure 2) are the lines between the azimuths of the summer solstice sunrise and the winter solstice sunset, or the azimuths of the summer solstice sunset and the winter solstice sunrise. We can see from Figure 2 that the cross of the cathedral is slightly off alignment with both these axes. It should be noted that such an alignment means that *no* sunset for any calendar date could ever align with the cathedral's west entrance. Indeed the cathedral's west door is about

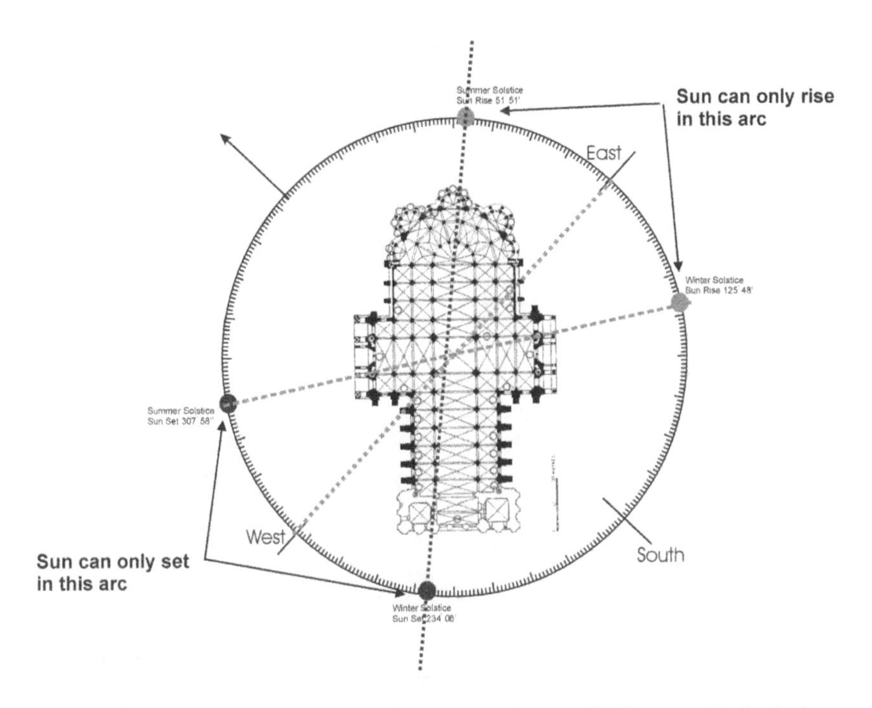

Figure 2: The apparent sunrise and sunset arcs around Chartres Cathedral. The long dotted line running length-ways through the building is the axis of the summer solstice sunrise (top) to winter solstice sunset (bottom). The shortest dotted line is the other solstice axis of winter solstice sunrise (right) to summer solstice sunset (left). The Equinox axis is marked by a dotted line running east to west and it can be seen that the transept crossing is aligned to this axis.

7° degrees further around to the south of the most southerly position of the winter solstice sunset for, according to the plan of the cathedral, the western door's azimuth is 227° and the azimuth of the winter solstice sunset is 234°.

The cathedral is set on a natural mound and in the twelfth century it overlooked the walled town. Additionally, for the latitude of Chartres, as the sun rises or sets on the date of the solstices, it cuts the line of the horizon at an angle of 42.5°, unlike more tropical latitudes where this angle is much closer to 90°.

Bearing this in mind, Figure 3 shows that in the evening, for a few days either side of the winter solstice, the orb of the sun will physically pass in front of the west door on its way to disappearing below the horizon. Furthermore, at the moment that the setting sun is aligned with the western doorway (azimuth of 227°), its elevation will only be 4°10'

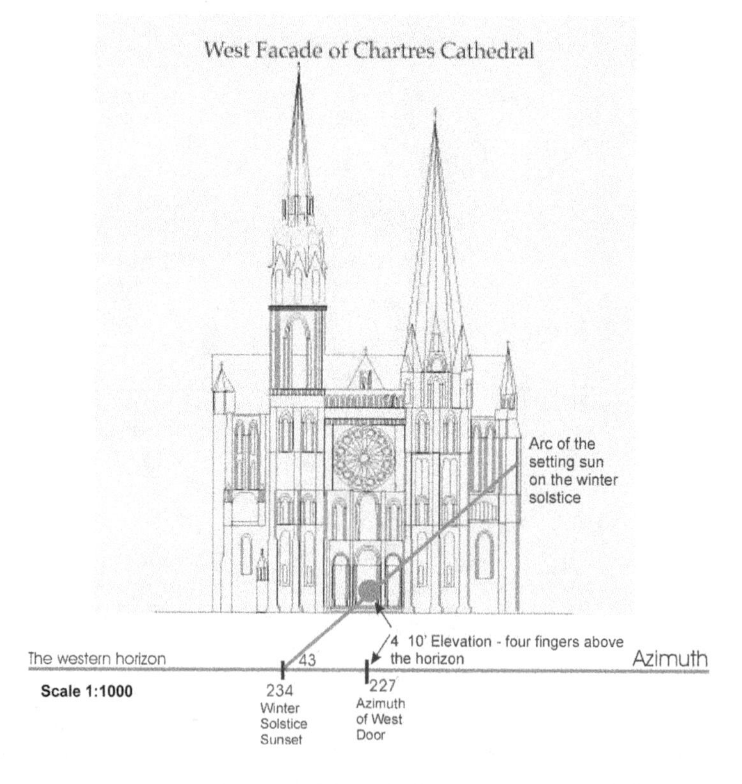

Figure 3: Chartres Cathedral and the winter solstice sunset (solid line) showing the arc of the setting sun and how it moves across the western door at sufficient elevation to be visible inside the cathedral. (Size and position of the sun in the door is approximate).

above the horizon. This elevation is about the width of three or four fingers held against the line of the horizon and such an elevation would be sufficient to allow the sun to be visible above the door's entrance steps and to be seen through the central doorway.

If the builders of Chartres had aligned the cathedral with the position of the setting sun on the horizon line, then the western steps would block the sun from view from the observer in the cathedral. However, the cathedral's alignment is orientated so that the western doorway faced seven degrees further south, and the observer would, therefore, be able to see the spectacle of the sun standing in the doorway on the day of the winter solstice.

Whether this is by design or by chance, the symbolism is significant. Recognizing that the building was designed with Christian Platonist theology in mind, then to allow the sun to enter the cathedral on the day of the winter solstice, the darkest day of the year, which is close to the day celebrated as the birth of Christ would seem too elegant to be a simple coincidence. Moreover, the sculpture in this centre doorway of the Royal Portal is that of the second coming of Christ. This theme of the Return of Christ is also supported in the West Rose located above the Royal Portal. Bearing in mind that the cathedral's orientation is older than the Royal Portal and possibly older than Christianity, this orientation with *any* building or structure on the mound of Chartres would more than likely allow the sun to 'enter' on 'the day of its birth,' or the day which is the beginning of the return of the light.

The transept solstice axis

The position of the winter solstice sunrise and that of the summer solstice sunset provides another solstice axis which mirrors the shorter transept axis of the cathedral (see Figure 2). According to the floor plan of the cathedral, the southern door has an azimuth of 138°, an orientation that allows the orb of the rising winter solstice sun to be visible in this doorway (see Figure 4). This doorway was the principal entrance to the cathedral in the medieval period and remained so until the advent of the railway when pilgrims started to approach the cathedral from the west.[36]

Thus, the south door also appears to have a link to the sun on the day of the winter solstice. For, on Christmas morning and a few days either side of Christmas, the orb of the rising sun would be seen to stand in the southern doorway at about 8.30 am local time and thus be seen to 'enter' the cathedral. Later that same day, at about 3.30 pm local time, the sun would be seen standing in the western doorway (Figure 2). It should be noted that this would be the only time of the year, weather permitting, when the orb of the sun could be seen to 'enter' the cathedral. This solar phenomenon suggests, therefore, a possible explanation for the unusual alignment of 47° north-east of the cathedral.

Since the orientation of the cathedral is far older than the present-day structure we will probably never know if this was understood or appreciated by the cathedral's twelfth-century architects; but, whether by design or accident, this orientation is strongly sympathetic to the neo-Platonist Christian theology of the light coming from the divine darkness.

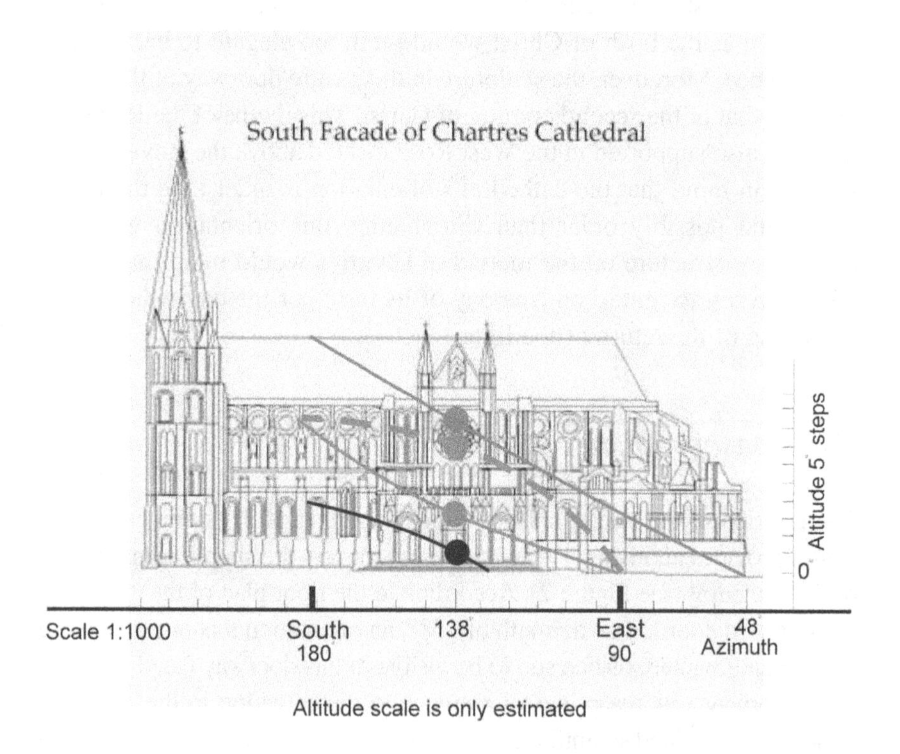

Figure 4: The south-eastern sunrise side of Chartres Cathedral. The elevations of the sun are only estimated with regard to their precise location against the cathedral (error of +/– 2°). The lowest line is the winter solstice sunrise, the top line is the summer solstice sunrise, the solid middle line is the spring equinox sunrise and the dotted line is the autumn equinox sunrise.

Equinox axis

The equinox axis is formed by the two positions at which the sun will rise and set on the days of both the spring and the autumn equinox and, as can be seen from Figure 2, the cathedral's transept crossing is aligned with this axis. Indeed, the four points of the transept crossing are aligned to the four cardinal points. This alignment may have been established by the time of the fourth-century Gallo-Roman chapel, but it is quite feasible to believe that the builders of the current cathedral intentionally incorporated this important alignment. Furthermore, this alignment gives another hint for the unusual orientation of the cathedral of 47° north-east. For it is only at the latitude of Chartres that one can gain this combination of having both the transept crossing aligned to the four cardinal points, while also being able to receive orb of the winter solstice sun in both the southern and western doors.

Conclusion

Chartres has long been proclaimed one of the great cultural centres of western spiritual and esoteric thought and contained within this great structure is a symphony of numbers, geometry, stone, glass and light which, I believe, can never be fully understood. This paper has, therefore, focused on the Christian theology of the twelfth century, its links with the colours of the twelfth and thirteenth-century glass and its effect on the human eye. Through this functional exploration of the twelfth-century lighting 'trick' we are lead to an appreciation of the importance of light and dark to the spiritual message of the cathedral. This, in turn, suggests a potential explanation for the persistence, since at least the fourth-century Gallo-Roman chapel, of all buildings on this mound maintaining the unusual alignment of 47° north-east — an alignment that allows the orb of the sun to stand in the southern (main door) and western doorways on Christmas day, the most sacred — and also one of the darkest —days of the year. Von Simson considers Chartres Cathedral to be the foundation stone of Christian Platonic tradition and one of the sources of Christian metaphysics.[37] With the timeless rhythm of sunrise and sunset illuminating its divine darkness it is still, after eight hundred years, radiating its empyrean message. It is a sacred space which actively encourages exploration and at the end of one's journey of exploration one muses that any answers, solutions or hypotheses that arise only serve to draw the researcher ever deeper into the labyrinthine mystery that is Christian Platonism.

> There is in God (some say)
> A deep, but dazzling darkness ...[38]

Notes

1 Eliot, 'East Coker,' from *The Four Quartets*
2 Charpentier, *The Mysteries of Chartres Cathedral*, p. 25.
3 Strachan, *Chartres: Sacred Geometry, Sacred Space*, p. 9.
4 Miller, *Chartres Cathedral*, p. 5.
5 James, *Chartres: The Masons Who Built a Legend*, pp. 3, 55.
6 Miller, *Chartres Cathedral*, p. 12; Parry, *Great Gothic Cathedrals of France*, p. 65; von Simson, *The Gothic Cathedral*, p. 164.
7 Strachan, *Chartres: Sacred Geometry, Sacred Space*, p. 9.
8 Parker, 'Letter Vof Dionysius the Areopagite' (1897).
9 Lilley, *Prayer in Christian Theology*, p. 20.
10 Blackburn, *Oxford Dictionary of Philosophy*, p. 29.
11 Parry, *Great Gothic Cathedrals of France*, p. 64.
12 Abbott Suger, *Abbott Suger on the Abbey Church of St Denis and its Art Treasures*, p. 63.
13 Strachan, *Chartres: Sacred Geometry, Sacred Space*, p. 25.
14 Strachan, *Chartres: Sacred Geometry, Sacred Space*, p. 31.
15 Strachan, *Chartres: Sacred Geometry, Sacred Space*, p. 40; von Simson, p. 29.
16 James, *Chartres: The Masons Who Built a Legend*, p. 39.
17 Strachan, *Chartres: Sacred Geometry, Sacred Space*, p. 50.
18 von Simson, *The Gothic Cathedral*, p. 31.
19 Strachan, *Chartres: Sacred Geometry, Sacred Space*, pp. 96, 97.
20 von Simson, *The Gothic Cathedral*, pp. 32, 35.
21 von Simson, *The Gothic Cathedral*, p. 51.
22 James, *Chartres: The Masons Who Built a Legend*, p. 139.
23 Miller, *Chartres Cathedral*, p. 18.
24 Johnson, *The Radiance of Chartres: studies in the early stained glass of the cathedral*, p. 7.
25 Johnson, *The Radiance of Chartres: studies in the early stained glass of the cathedral*, p. 9.
26 Johnson, *The Radiance of Chartres: studies in the early stained glass of the cathedral*, p. 10.
27 Johnson, *The Radiance of Chartres: studies in the early stained glass of the cathedral*, p. 22.

28 Johnson, *The Radiance of Chartres: studies in the early stained glass of the cathedral*, p. 24.
29 James, *Chartres: The Masons Who Built a Legend*, p. 178.
30 Johnson, *The Radiance of Chartres: studies in the early stained glass of the cathedral*, p. 24.
31 Johnson, *The Radiance of Chartres: studies in the early stained glass of the cathedral*, p. 63.
32 Johnson, *The Radiance of Chartres: studies in the early stained glass of the cathedral*, p. 23.
33 Johnson, *The Radiance of Chartres: studies in the early stained glass of the cathedral*, p. 41.
34 James, *Chartres: The Masons Who Built a Legend* p. 90.
35 Strachan, *Chartres: Sacred Geometry, Sacred Space*, p. 12.
36 Parry, *Great Gothic Cathedrals of France*, p. 76.
37 von Simson, *The Gothic Cathedral*, p. 39.
38 Henry Vaughan, 'The Night.'

Bibliography

Aquinas, Thomas, *St. Thomas Aquinas on Politics and Ethics* (New York: Norton and Co. Inc., 1988).

Blackburn, Simon, *Oxford Dictionary of Philosophy* (New York: Oxford University Press, 1996).

Eliot, T. S., 'East Coker,' in *The Four Quartets* (New York: Harcourt Inc., 1943), pp. 23–34.

Charpentier, Louise, *The Mysteries of Chartres Cathedral* (New York: Avon Books, 1972).

James, John, *Chartres: The Masons Who Built a Legend* (London: Routledge & Kegan Paul, 1982).

Johnson, James Rosser, *The Radiance of Chartres: studies in the early stained glass of the cathedral* (New York: Random House, 1965).

Lilley, A. L, *Prayer in Christian Theology* (London: Student Christian Movement, 1925).

Miller, Malcolm, *Chartres Cathedral* (Hampshire: Pitkin Guides, 1985).

Parker, Rev. John, 'Letter Vof Dionysius the Areopagite' (1897), at http://www.voskrese.info/spl/index.html (accessed 26 March, 2004).

Parry, Stan, *Great Gothic Cathedrals of France* (New York: Penguin Putnam Inc., 2001). Ruth-Heffelbower, Duane, 'Toward a Christian Theology of Church and Society as it Relates to Restorative Justice' (paper presented at the 4th Annual Restorative Justice Conference, 25 October 1996, Fresno, California, USA) at http://www.fresno.edu/pacs/docs/speech.html.

Strachan, Gordon, *Chartres: Sacred Geometry, Sacred Space* (Edinburgh: Floris Books, 2003).

Suger, Abbott, *Abbott Suger on the Abbey Church of St Denis and its Art Treasures,* Translated by E. Panofsky (USA: Princeton, 1979).

Vaughan, Henry, 'The Night,' in *The Poems of Henry Vaughan,* volume 1, E. K. Chambers, ed. (London: Lawrence and Bullen Ltd., 1896) pp. 251–53.
von Simson, Otto, *The Gothic Cathedral* (New York: Pantheon Books Inc., 1956).

4. Life Across the Cosmos

NEVILLE BROWN

This chapter discusses the Anthropic Principle, *which, in its various forms, has been developed to account for the advent, tenacity and adaptability of life on a relatively salubrious planet earth. The paper considers the implications of evolution, the notion of panspermia and the patterns it might assume. It discusses the limited evidence for its occurrence and the possibility that cosmic macrostructures may be conducive to it.*

Our homely Earth

The first thing I would like to say about the way professionals look at the question of Life across the universe is that there is a long-standing tradition (which in the last fifteen years or so has enjoyed a considerable revival) of talking about the *Anthropic Principle* — the 'anthropic cosmological principle.' Strictly speaking, 'anthropic' means human-friendly; and my understanding is that this is what it did connote when first used in this context around 1930 by a British theologian called J. R. Tennant. As time has gone by, however, the question has become whether the universe is well disposed to life in general.

Opinion is divided over whether one endorses a strong anthropic principle or a weak one or, of course, no such principle at all. Still, the weak versus strong dichotomy has to be too simplistic. In fact, when you take the respective definitions together, they amount to a distinction without a difference. Thus, a weak anthropic principle is understood to refer to a universe with a few niches (estimates are never given) in which life could form. A strong anthropic principle is duly taken to be one where life occurs more generally.

Let us take, first of all, a precept of which we are all well aware. This is that, however uncomfortable we may make it by certain aspects of our own behaviour, the little planet we live on is basically very life-friendly: the widespread availability of water, of oxygen and carbon, and so on.

Moreover, these last ten years or so, other attributes have come to our notice which strengthen this claim. Three which I have encountered are:

1. The fact that our Moon is biggish in relation to its planet gives it considerable leverage in various respects. Thus it limits secular variations in the inclination of the Earth's axis; and this in turn helps to limit temperature fluctuations, season to season or era to era.
2. Next there are the impacts on the Earth of meteorites and comets. Their recurrence is now convincingly portrayed as posing a serious threat. Nevertheless, it is less so than it might be, but for the gravitational deflections effected by mighty Jupiter.
3. The third point is that we have become vividly aware of just how tenacious life can be. If one has a quarter of an acre of back garden, one can observe plenty of tenacity in one form or another. More systematic studies have underlined this truism.

In addition, what particularly strikes me is the evidence being amassed about life around the deep ocean vents. Due to plate tectonics, volcanically active ridges are found longitudinally down the mid-Atlantic and mid-Pacific, as well as in other deep-water locales. Down there, many thousands of feet below sea level, life develops quite extensively. Furthermore, samples recovered from the mid-Atlantic ridge-line are shown to be genetically related to forms near or on the surface. How did life get to, or evolve within, such depths? How does it survive without free oxygen? How does it tolerate the heat and the overpressure of water?

An accommodating universe

Testing the Anthropic Principle is fraught with disputation about whether the universe itself could have originated and developed had it been designed even slightly differently from how it was. On this score, astronomers come up with statistics that are incredible in two contrary ways. On one hand, one cannot easily believe that they can do calculations as definitively as they attempt. On the other hand, some of the results could be out by some orders of magnitude and still be quite amazing.

Take the calculations concerned with the future expansion or contraction of the universe. The argument has been that, were it expanding slightly faster than it is doing, it would be destined to disperse continuously and arrive, many billions of years into the future, at a state of formless heat death. Were it currently expanding slightly more slowly, however, it would be destined to collapse in upon itself, perhaps perpetuating thereby an endless cycle of expansion from one Big Bang which eventually gives way to contraction into the next one. Some years ago, John Barrow, Professor of Astronomy at Cambridge and highly regarded by fellow professionals, calculated that the universe could not be poised thus if its rate of expansion were different in either direction by one part in 10^{30}: one part in a one with thirty noughts after it, up to the decimal point.

In the interim, however, several germane facts have been discovered. Thus, it has now been confirmed that the zillions upon zillions of neutrinos which permeate everything do have a rest mass; and that a mysterious 'dark matter' occurs quite abundantly. Then again, evidence collated several years ago shows that, contrary to previous 'common sense' assumptions, the rate of expansion of the cosmos has accelerated these last several billion years. Other examples of pretty fine tuning as calculated could include the balance between matter and anti-matter; and the ratio between the mass of a proton and that of a neutron. Still, even if the balance between expansion and contraction was critically poised to, say, only one part in a hundred, this would still represent a subtle finesse.

Two years ago, I would have been almost the last person to say, 'Well, there must have been an unmoved mover. There must have been a mind behind all this.' Today I am more open. Evidently there is much sophistication in the programme out of which the universe has developed. However, what philosophical conclusions we draw from this in terms of our situation as human mortals is another question again.

Opportunistic life

Looking now at how life has asserted itself in the universe in the way it appears to have done, we do well to work from the premise that planets are far and away the most likely venues for life. There are dust clouds in interstellar space, but the dust is too thin and disaggregated to harbour water or otherwise support articulated life. Somewhat similar arguments apply to the stars. Their significance therefore lies in generating the equatorial discs out of which planets may coalesce. Mathematical modelling predicts both the discs and their eventual consolidation.

In 1995, the first planetary 'blob' was observed. You have to put it like that because one could not prove consolidation had yet taken place, a detectable aggregation of material notwithstanding. The said discovery took place not through direct visual sighting but by observing how the parent star was moving with an irregularity which suggested something fairly big was periodically close by.

Since then, over a hundred such blobs have been identified. Nevertheless, this technique has its limitations. All else apart, you are not likely to discern a planet, actual or prospective, much smaller than Jupiter. Coming into service now are ultra-modern infra-red telescopes able to focus on the longer-wave, lower temperature radiation associated with that part of the electromagnetic spectrum through which a cool planet stands a real chance of being optically distinguishable from its hot parent star.

So far, so good. But one has to be more cautious when using our own planet as a test case for the evolution of life. We can all take the point that our Earth seems to be quite a niche of life-friendliness. Nor does life appear to have taken too long a time to get going here. What does seem to have taken inordinately long, however, is for life to move to rather higher levels of complexity.

Let us try and identify a few milestones. The Earth consolidated out of being a mere blob about 4,500 million years ago. For a long time thereafter it was subject to cometary bombardment of the order of a thousand times heavier than what we observe from later, more familiar, geological eras. Then some 3,800 million years ago, this impaction diminished sharply. Fairly soon, methane-based life appeared. I am not at all sure just what 'fairly soon' means in this context. Nor do I know that anybody else is. A few hundred thousand years or a few million? At all events, we do mean a concurrence a good deal closer than a few hundred million years. Yet, it then takes well over a billion more years for life to transmute to oxygen/carbon dioxide dependence, a shift which entailed a huge crisis in Gaia terms. One upshot was an atmosphere that remained considerably oxygen-rich; and which has been a crucial part of an enduring oxygen circle. Measuring the atmospheric presence hundreds of millions of years ago at all accurately is a tricky business. Still, the tentative indications are that the oxygen percentage may have stayed close to twenty per cent ever since this transition.

The said methane-oxygen adjustment took place some two billion years ago. The world then goes through another one-and-a-half billion years (characterized by, it would seem, tiny creatures in the bacteria/

amoeba/worm range) before the first signs of hard skeletons appear. All of which does suggest that life around the universe today could be considerably microbial, a circumstance which would explain our failure to date to make contact.

The late Stephen Jay Gould, a well-regarded evolution theorist, made a judgement which his colleague, friend and critic Richard Dawkins, saw as quite off-key. It was that evolution proceeded in fits and starts — in other words, it was 'punctuated.' This notion is well in tune with that of evolution. In one of his last essays, Gould remarked on how we flatter ourselves that we are in the great age of Man or the great age of the mammals or something. But if you look at the total biomass of life forms on this planet, the truth probably is that we are still in the age of the microbes, certainly 'if the deep hot biosphere of bacteria within sub-surface rocks matches the upper estimates for spread and abundance.'[1]

Panspermia

Related to this, but more so to the question of the origination of life as a whole, is the debate about whether, when life first got going on our Earth, this was due (or likely due) to seeding from afar. About thirty years ago, this Panspermia question was formulated —or, if you like, revived - by the late Sir Fred Hoyle and Professor Chandra Wickramasinghe, currently head of the Centre for Astrobiology at Cardiff University. Their contention has been that 'the difficulty with which life forms, even if all the ingredients are there, is so immense that, if life gets going anywhere, presumably within our galaxy, it has got to be able to seed around.' They came up with one calculation that was particularly remarkable. They concluded that if you had a full assortment of 250 life-related proteins and you were 'cooking' them to become proactive and form life, the chances of their actually achieving this one particular afternoon were one in ten to 5,120. It is not revealed to me how they arrived at this figure as opposed to one in ten to 5,119. Never mind, the reckoning commands attention.[2]

Even more arresting, because more visual, was this illustration by Hoyle. Suppose you went to a metalworkers' giant scrapyard (broken up cars and so on). Then imagine everything was thrown into the air and a proportion came down as a perfectly assembled 747. The chances of this actually transpiring would be greater, he surmised, than those of life emerging spontaneously on a given occasion.

The Panspermia thesis made quite a lot of headway, but much will ultimately depend on what evidence is forthcoming. What is available to date is thin yet intriguing. One or two thousand tonnes of dust hit this Earth every day. About one per cent is travelling especially fast — that is, more than 100 km per second. This probably means it has come from outside the solar system. Then again, comets have a capacity to pulse out organic materials out as they zoom past us. This is what Halley's comet did during its 1986 visitation. Also the composition of a meteorite found in Antarctica indicates it is of Martian origin; and it does bear some well-ordered structures which could derive from life forms. So modest a data base is by no means proof conclusive of Panspermia, but it does suggest that the question is worth pursuing.

A further dimension in this dialectic is whether the macrostructure of the universe could allow Panspermia. As is well known, the cosmos is organized into vast assemblages of stars known as galaxies. These tend to be perhaps several hundred thousand light years across, and separated from their neighbours by one or two million light years. There are clusters of galaxies and, indeed, superclusters of galaxies. The total number of galaxies appears to run into tens of billions.

The Milky Way is, of course, *our* galaxy. Our local experience is that its stars (of which our sun is one) tend to be several light years away from each other.[3] With about 100 billion stars altogether, it probably counts as quite a large galaxy. Subject to further consideration, this cosmic arrangement seems to make Panspermia feasible but by no means easy. A weak Anthropic Principle, perhaps?

Notes

1 Gould, Stephen Jay, *The Structure of Evolutionary Theory* (Harvard: Belknap Press, 2002), p. 898.

2 Hoyle, Fred and Chandra Wickramasinghe, 'Panspermia 2000,' in their *Astronomical Origins of Life* (Dordrecht: Kluwer, 2000), pp. 1–17.

3 A 'light year' is the distance electromagnetic radiation travels in a year. It is in excess of five million, million miles.

5. Imagining Eternity: Weaving 'The Heavens' Embroidered Cloths'

JULES CASHFORD

This talk will explore the idea that the sky has always offered the human psyche an image of eternity, serving as a heavenly screen on which to play out the drama of our eternal selves —firstly through the Moon, and then the Sun and the pattern of the stars, and subsequently through the invisible world beyond the senses, known in Christian symbolism as the 'kingdom of heaven.' This journey of the psyche, which has been a movement away from an original ensouled reality on Earth, may have reached its farthest point, making possible a new relationship to the cosmos through the Imagination, and so weaving a new story of a re-enchanted universe in which eternity again enters into time. This will be explored primarily through humanity's myths and images of the Moon.

In Yeats' poem 'He Wishes for the Cloths of Heaven,' the author writes:

> Had I the heavens' embroidered cloths,
> Enwrought with golden and silver light,
> The blue and the dim and the dark cloths
> Of light and night and the half-light,
> I would spread the cloths under your feet.
> But I, being poor, have only my dreams.
> I have spread my dreams under your feet.
> Tread softly because you tread on my dreams.[1]

It is our dreams which fill the sky with the infinite forms of our desires and longings, feeling it to be beyond all earthly limitation, and so set free for our imagining. It seems that this sense of the infinite above us imperceptibly infuses what we simply see with what we deeply value, turning the sky into a heaven, becoming both the place *beyond* the sky and a *locus* of value, the region of immortal beings: the goddesses and gods of earlier times and, later, God and his Angels. On a clear day you can see for ever.

Yet these words in origin are practical: Our word for 'sky' comes from the Old Norse for cloud and shadow, and 'heaven' from Old English as that which is heaved, lifted, raised up. This meant that the visible sky was often taken to be the floor of heaven, seen from underneath, as in *The Merchant of Venice*, where stars take the place of clouds:

> Sit, Jessica. Look how the floor of heaven
> Is thick inlaid with patines of bright gold.
> There's not the smallest orb which thou behold'st
> But in his motion like an angel sings,
> Still quiring to the young-eyed cherubims;
> Such harmony is in immortal souls;
> But while this muddy vesture of decay
> Doth grossly clothe it in, we cannot hear it.[2]

But we may hear its echo in the tales that come down to us about the harmony of the universe, which is to say in our myths, in poems, images and visions, and even in our charts. As Yeats knows, it is our dreams, poor as they may be, which embroider the cloths of heaven, just as the cloths of heaven show us back our dreams. For Coleridge, gazing at the Moon in the harbour of Valletta in Malta, Nature offers us the language in which to think about ourselves:

> In looking at the objects of Nature while I am thinking, as at yonder moon dim-glimmering thro' the dewy window-pane, I seem rather to be seeking, as it were *asking*, a symbolical language for something within me that already and forever exists, than observing anything new.[3]

Myth, for Joseph Campbell, is 'the secret opening through which the inexhaustible energies of the cosmos pour into human cultural manifestation,' and for Jung it 'reveals the nature of the soul.'[4] He continues:

> All the mythological processes of nature, such as summer and winter, the phases of the moon, the rainy seasons, and so forth, are in no sense allegories of these objective occurrences; rather they are symbolic expressions of the inner, unconscious drama of the psyche which becomes

> accessible to man's consciousness by way of projection — that is, mirrored in the events of nature. The projection is so fundamental that it has taken several thousand years of civilization to detach it in some measure from its outer object.[5]

This process of mirroring is mutual, however, for if the heavens are within us and without us, then sky and psyche reflect each other endlessly. Origen, in the third century AD, wrote: 'Understand that you are another little world, and have within you the sun, the moon and also the stars.'[6] But which Sun and Moon, which sky? Our perception of them changes over time, and we cannot talk as if we know which 'the real one' is.

Images of eternity

One of the inner unconscious dramas which we can see played out in the infinite mirror of the heavens is the drama of what we might call eternity, our 'immortal longings' and eternal selves or soul, the dimension of the psyche that believes only in life, understanding death as a change of form. The location of this image of the eternal has also changed over time, but its first manifestation was the Moon.

Ancient watchers of the night skies saw a figure of light who was always changing in a way that was always the same. Perpetually moving — from crescent to full to crescent to dark and again to crescent — the Moon tells one fundamental story: birth, growth, fullness, decay, death and rebirth. It is the story of transformation. Early people perceived the Moon's waxing and waning as the growing and dying of a celestial being, whose death was followed by its own resurrection as the New Moon. The essential paradox, which leads the mind beyond the frame of the senses, is that the cycle, the whole, is invisible yet contains the visible phases, as though the visible comes out of and falls back into the invisible — like being born and dying, and being born again. The instinctive identification of the people with their Moon meant that they interpreted the Moon's rebirth as offering a similar promise for human beings in their own waning and death.

Early human beings saw their own lives reflected in the life of the Moon. The drama of the Moon was the human drama magnified, given dignity and solemnity, and brought into the cosmic drama of all creation.

It was different with the Sun. Though it also travelled through the black ocean of the underworld, and later fought and overcame the demon of darkness, it did not lose all its colour or shape or the whole of its light, it did not grow old. The Moon, by contrast, suffered painfully the gradual but relentless loss of its own body of light. Just as it had reached its culmination at the full, its shape diminished, its light dimmed, first imperceptibly and then irrevocably, until it was nothing at all, 'mere oblivion:' '*Sans* teeth, *sans* eyes, *sans* taste, *sans* everything,' as Shakespeare has it in *As You like it*.[7] Waning was like ageing. The Moon was like humanity, both caught in the universal law of becoming which ends in death. Tragedy was the fate of the human and the *lunar* condition.

Yet the Moon was also *not* like humanity: it transformed the story of death into a story of rebirth. It brought forth a new Moon. In its perpetual return to its own beginnings, the Moon unified what was broken asunder, and so the Moon held out a promise that death was not final, only a change of form. In this way, duality, made visible as the waxing and waning Moon, was both embodied and transcended in the ever-recurring cycle. The paradox arises that, by dying, the Moon is reborn; by going away, it eternally returns, 'time and again.' So if, with the awareness of time came the longing for eternity, then the Moon would seem to answer the need it originally provoked. It appeared, in its acts of measuring, to create time, and then, month by month, to redeem it. So, in a sense, time *and* eternity come into being together through the Moon. 'May the gods accord me a life renewed each month, like that of the Moon,' the Babylonians used to say.[8] So the Moon became a visible symbol of hope, the light that shone in the darkness of the human psyche.

Stories and images from around the world suggest that most, if not all, ancient cultures passed through a stage of interpreting essential aspects of their reality through the Moon. The perpetual drama of the Moon's phases became a model for contemplating a pattern in human, animal and vegetable life, one which did not just include the idea of life beyond death, but, more vitally, took its inspiration from the Moon's miraculous rebirth, its eternal being. From a broad perspective, the Moon was one of many expressions of the culture of the Great Mother Goddess, in which Earth and Moon were understood as one manifestation in dual aspect. But the pattern of eternal return was more immediate, easier to see and comprehend, than the longer patterns of the year and the cyclical seasons of Mother Earth. Perhaps for this reason the powers attributed to the Moon are so astonishing — powers over birth, fecundity, growth, destiny, death and rebirth and, of course, time which came within the embrace of eternity, not in opposition to it: time as 'the moving image of eternity,' as Plato puts it in *The Timaeus*.[9]

Time was once measured everywhere by the phases of the Moon. Or perhaps we could say that the notion of time came into being when the phases of the Moon fell into a cyclical pattern and became measurable. The dual rhythm of constancy and change is also the beginning of 'measurement,' a word with the same etymological root as 'Moon' in all Indo-European languages. For, while the course of one day can be marked by the Sun, the earliest tracking of time longer than a day was only possible through the Moon.

But this was 'time' as a physical phenomenon, something lived continuously from night to night and then from phase to phase, and so from Moon to Moon. Measurement is now a quantitative term — how many days in a month? Yet the word 'measure' still retains its older qualitative meaning in such phrases as 'in due measure,' and 'love is the measure of all things.' Ancient measurement contained both meanings: measuring time by following the phases of the Moon, and also taking time itself as the measure, in the sense of reckoning the mode of behaviour due to the time, which was originally the particular phase of the Moon. This was Time as Quality. So the Moon's story of birth, death and rebirth becomes the story of time which is lived by human beings on Earth.

We can see this in the Palaeolithic Goddess of Laussel, one of the oldest sculptures in the world. She comes from the Dordogne in France, a few miles from the great cave at Lascaux and was made around 20,000 BCE.

On a ledge overhanging a high rock shelter with a sheer drop down to the ground, the reclining figure of a pregnant woman once soared up out of pink limestone rock. Paleolithic sculptors had chiselled her with tools of flint out of the soft limestone and given her to hold the horn of a bison, crescent-shaped like the Moon, and sprinkled her with red ochre, the colour of blood. In her right hand she holds up the crescent horn of the bison, which is clearly incised with thirteen downward strokes, while her left hand rests lightly on her swelling belly, which, sloping towards the fertile valley below, would have commanded the whole landscape. Her head inclines towards the crescent-shaped horn, as though contemplating the mystery held in the hollow of her hand, to which her other hand, resting on her pregnant womb, also bears witness.

The thirteen strokes upon the crescent horn are unlikely to be random, as the work of Alexander Marshack, in his book *The Roots of Civilization*, makes clear.[10] Thirteen is the number of days from the birth of the first crescent to just before the days of the full Moon — the period

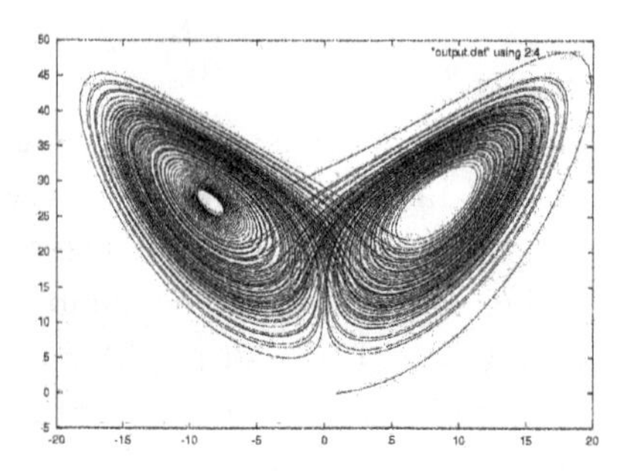

Figure 1: Goddess of Laussel. Limestone bas-relief. Upper Perigordian, c. *22,000–18,000* BCE, *Height 17 in. (43 cm), Dordogne, France (Musée d'Aquitaine, Bordeaux).*

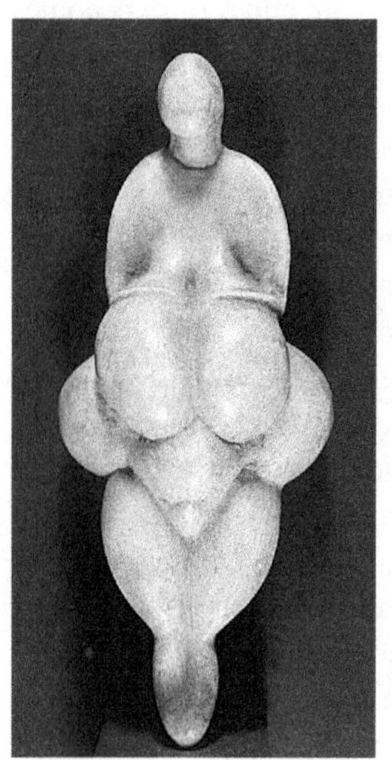

Figure 2: Goddess of Lespugue, front and back. Mammoth ivory statue, c. *20,000–18,000* BCE. *Height 5 ½ in. (14 cm), Haute-Garonne, France (Musée de l'Homme, Paris).*

of waxing; thirteen is also the number of cycles that may make up an observational lunar year (that is, a solar year of 365 days measured by the number of Moons). Through the movement of the sculpture and the echoing of the motif of notched strokes on the horn and the hand, the artist creates a relationship between the growing of light in the Moon, the growing of the child in the womb, and also, perhaps, from the placing of the sculpture as the summit of the landscape, the growing of vegetation in the womb of Earth. Whether this is the Great Mother Goddess, pregnant with the world in her waxing mode, or whether she is the woman whose cycles of fertility are governed by the Moon, an essential accord between celestial and earthly orders is in either case celebrated. A more eloquent testimony to the unity of heaven and earth would be hard to find; 'As above, so below' — as it was phrased in the Emerald Tablets of Hermes Trismegistos, 20,000 years later.

The Moon and living time

What a later age would distinguish as two separate tales — the Moon making time measurable and the Moon making the growing and diminishing of things in an eternal rhythm — was to an earlier age one comprehensive story of life in time: living time. Because the Moon, returning to its own beginning, created a cycle, so, analogously, it became the *ruler* of everything else that was cyclical — the ebb and flow of seas, the nightly dews, the seasonal rains, the flux of rivers and tides of blood. These are the essential waters of life that come and go — in a day, a month, a year and a lifetime. As the cycle of the Moon appeared to correspond so precisely to the monthly rhythms of the wombs of women, it is likely that the first reckoning of time was made by women calculating the timing of their menstrual cycles from Moon to Moon. In many languages the roots of the words for Moon, menstruation and mind are the same or related. Since the timing of birth could also be calculated by the Moon (as 10 Moons of 28 days), so the Moon was assumed to govern their phases of fertility, and, by extension, the creative cycles in all living beings.

The very pregnant Goddess of Lespugue from 22,000 BCE (Figure 2), who lay buried for millennia in a muddy ditch, has ten vertical lines incised from waist to buttocks, suggestive of the ten Moons from conception to birth. But, again, this is not simply linear numerical time; it is time as quality and substance, cyclical time, continually reborn. For

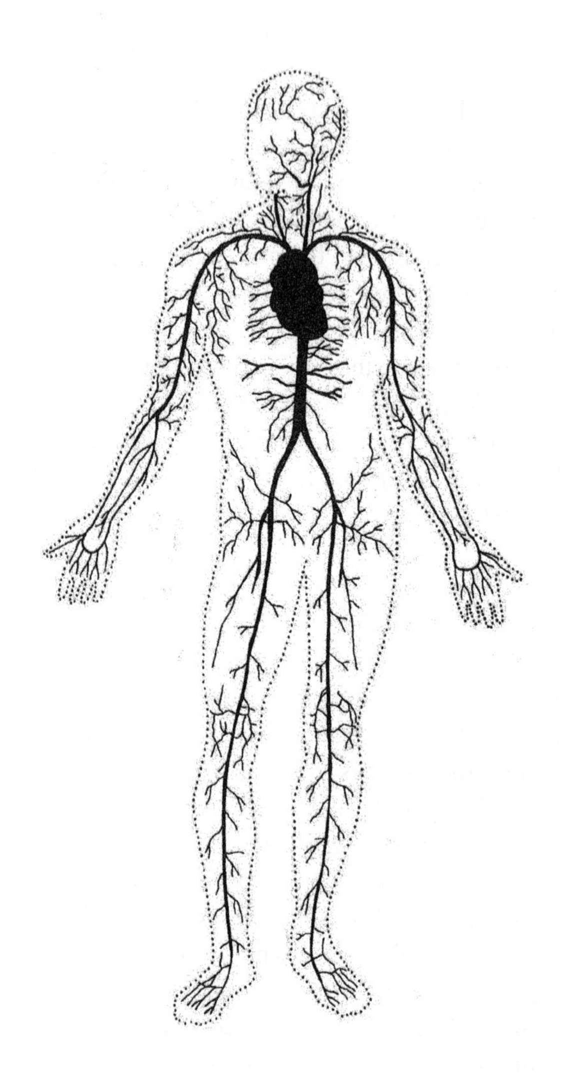

Figure 3: Osiris-aah (aah meaning 'Moon'), with Crescent and Full Moon upon his head, holding the Djed column, symbol of resurrection, with flail and sceptre as signs of his power. Temple of Seti I, Abydos, c. *1300* BCE.

the Moon appeared to give birth to itself, and so was regarded as the primary source of fertility on Earth: the increase and decrease of animals and plants, and all the milk, semen, sap and moisture of growth. As with humans and animals, so with plants: the Moon was the origin of the rhythmic rising and falling of the sap, the seasons of sowing and reaping, the power of growing that came and went in a lunar rhythm.

Revealed in so many different phenomena, the Moon fosters a vision of the universe as a coherent pattern of relationships, all informed by the same laws which act upon each thing in a similar way. This sets up a system of analogies and correspondences between all levels of life, both visible and invisible. For though the Moon divides and orders temporal phenomena through its distinctive phases, it also unifies them in the larger perspective of its revolving cycles, and this creates reverberations, echoes and harmonies between one part of the whole and another — just as though the universe were one great cosmic web of relationship (as modern sub-atomic physics claims it is). In 'lunar thinking' there can be no part without the whole.

Eternity and time

In early myths, the Moon was referred to *both* as eternal *and* as the one who makes, measures and suffers time. There was no contradiction here since the eternal Moon, entering into time and also transcending it, may be thought to *transform* time.

In North America the Sioux Indians called the Moon 'the Old Woman Who Never Dies;' the Iroquois called her *Aatensic*, 'the Eternal One.'[11] For the Egyptians the Moon god, Thoth, was 'the maker of eternity and creator of everlastingness,' as well as the 'scribe of time.'[12]

In Egypt, Osiris was called 'Lord of Eternity,' and worshipped as the Moon from the oldest Pyramid Texts (*c.* 2500 BCE): 'You are born in your

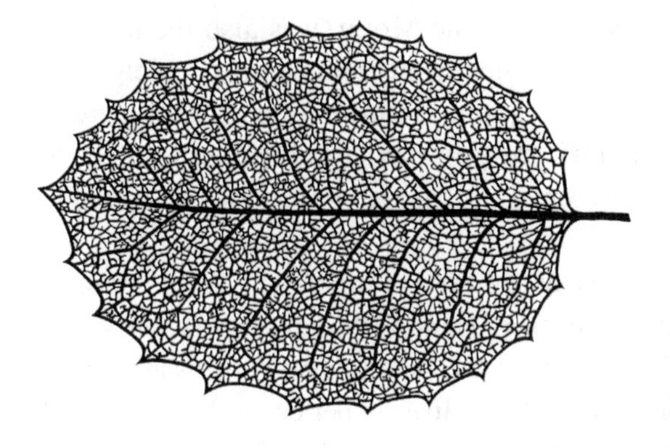

Figure 4: Cow-headed Isis watering the plants in a sacred lake, out of which rises a man-headed hawk as the soul of Osiris. The hieroglyphs read: 'This is the soul of Osiris speeding upwards.' Bas-relief, Ptolemaic Temple of Isis at Philae. From E. A. Wallis Budge, Osiris and the Egyptian Resurrection *(New York: Dover Publications, 1973), i, p. 8.*

months like the moon;' 'You appear at the New Moon.' Over a thousand years later, Rameses IV, in a hymn to Osiris, says to him: 'Lo, thou art the moon on high; thou becomest young at will and agest at will.' The story of his death and resurrection follows the cyclical pattern of the Moon, extended to vegetation and the Nile, the Moon's water on Earth. In the same hymn Ramses continues: 'Thou art the Nile...gods and men live from thy outflow.'[13]

In Mesopotamia the Sumerian-Babylonian Moon god, Nanna-Sin, was known as 'the fruit that gives birth to itself,' the Lord, 'ever-renewing himself,' who 'measures the days of a month,' 'Lord and Giver of Life,' Lord of Cattle and Plants.[14] Inanna was the daughter of the Moon god, Nanna-Sin, and eventually took over her father's role, becoming Queen of Heaven and Earth, 'begetting mother,' abiding 'within the Spirit.' She was 'the pure torch that flares in the sky, the heavenly light, shining bright like the day.'[15] She made the rain fall and the plants grow. In the drama *The Descent of Inanna*, she journeys from the Great Above to the Great Below to visit her dark sister, Ereshkigal, in the underworld of darkness. Disrobing her bright jewels of Heaven at seven gates, she hangs like a corpse on a hook for three nights (the number of nights the Moon is 'dead') until she is reborn and ascends into Heaven, gathering her shining garments around her as she goes.

In Latin inscriptions, the Moon was given the epithet of 'the eternal' and the crescent was carved on their tombstones above their names. In Vedic India, *Soma*, god of the Moon, was also the name of the *Soma* plant and the ambrosial nectar of immortality, which is the food of the gods and flows throughout creation as rain from heaven, water in streams, sap in plants, milk in cattle and semen in men; he is the First Husband of Women, Lord of the waters of life, animals and plants.

In China, the Moon hare pounds the herb of immortality in the crescent cup of the reborn Moon, from the tale of Heng'O who swallowed the herb of immortality on Earth and floated up to the Moon, upon which she vomited the herb which became the hare, and now lives in her Great Palace of Cold. Her husband, Hen Yi, following her, found himself in the Sun. Once a month at the Full Moon they meet for a Sacred Marriage of Yin and Yang.

In Polynesia, the Moon is perpetually renewed in the waters of *Tane*, the eternal source. In Hottentot and Bushman tales, the Moon would have given immortality to humans if the hare had not lied or squandered the chance.

Yet when the Moon grew thin and disappeared, it was evidently experienced to be a death as real as the death of human beings. The Nandi of East Africa called the twenty-eighth day: 'the Moon is nearing death'; the twenty-ninth: the people discuss whether the Moon is dead, or 'the Sun has murdered the Moon'; and the thirtieth, 'the Moon is dead' or 'the Moon's darkness.'[16]

All over the world, people danced to bring the Moon back to life. The North American Iroquois Indians danced, as they said, 'for the 'eternal one,' for the sake of her health, when she is sick.' The Incas of ancient Peru prayed: 'Mama Quilla, Mother Moon, do not die, lest we all perish.' In the nineteenth century, the old men among the Californian Indians would summon the young men to celebrate the resurrection of the Moon by dancing in a circle, chanting: 'As the Moon dieth, and cometh to life again, so we also having to die will again live.'[17]

At this early stage, when the specifically human consciousness could be said to be gradually emerging from the Collective Unconscious, it seems to have been the vision of the Moon's death and rebirth which carried people's reflection on death. Or rather, to talk in the language of projection, it was the *interpretation* of the Moon's disappearance and reappearance *as* death and rebirth which allowed these reflections to take place, and so gave the inner idea of rebirth an image.

Jung suggests that the idea of life after death is a primordial image, one that belongs to us as part of being human:

> Do we ever understand what we think? We understand only such thinking as is a mere equation and from which nothing comes out but what we have put in. That is the manner of working of the intellect. But beyond that there is a thinking in primordial images — in symbols that are older than historical man; which have been ingrained in him from earliest times, and, eternally living, outlasting all generations, still make up the groundwork of the human psyche. It is possible to live the fullest life only when we are in harmony with these symbols; wisdom is a return to them. It is a question neither of belief nor knowledge, but of the agreement of our thinking with the primordial images of the unconscious. They are the source of all our conscious thoughts, and one of these primordial images is the idea of life after death.[18]

The Moon and time

Because the Moon was originally a sacred being, it is inconceivable that the Moon was seen merely as a way of making time visible and measurable. We could imagine that, for the early mind, the Moon brought time into being in such a way that it also proposed the way to live in that time: a daily life in harmony with the laws of eternity. But, as Coleridge says in his poem 'The Ancient Mariner,' 'The moving moon goes up the sky and nowhere does abide.' The primitive way of stopping the flow of time was to focus on the dramatic moments of the Moon's cycle, which are the points of discontinuity where something new happens: the first glimpse of the New Moon in the Crescent, the perfect orb of the Full, the first loss of light in the Waning, and the total absence of light in the Dark Moon. These are the numinous moments of intensity, when time becomes transparent to its source, and eternity, as it seems, enters into time in the form of a story.

Ritualized time expresses the idea of the *quality* of time. For the time, say, of the New Moon is experienced as qualitatively different from any other time, with a unique character of its own — almost (to use a spatial term) in the manner of a *temenos*. In Greece, this *temenos* was the sacred space 'cut out' (the meaning of *temnein*) around the temple of the gods, which set that space apart from the rest of the land, and within which definite laws operated. Earthly life took its story from the Moon's story, so the New Moon in Heaven was also the New Moon on Earth, when anything new would prosper. Correspondingly, each ritual change of the Moon had its own sacred laws, and these were written into the laws of events on Earth. Earthly affairs are then bound to share in the quality of the time, as they are implicated in the character of the Moon at 'any one time.' What is remarkable is the *range* of implication, extending to everything that grows and dies, which on Earth is everything.

The way to live in this sacred world was naturally to put human endeavours into accord with the Moon's energies of increase and decrease, more immediately in order to share in their temporal benefits, but perhaps primarily to continue to participate in their powers of renewal in the life eternal.

Broadly, those things that were required to grow — sowing, planting, grafting, marrying, conceiving, giving birth, business enterprises — should be timed for the waxing. The folk custom of jingling the silver in your pocket at New Moon, so your wealth would grow with the growing crescent, has its origin in this rite. Those things that you wanted to diminish — pains, the sap of a cut branch, anything that would bring benefit by decrease — should be timed for the waning, such as felling timber, harvesting crops, mowing the grass, washing linen and treating warts. At its most general, New Moons were for beginnings, marrying, having babies, divining the future. Plutarch writes that the Athenians chose the days near the conjunction of Sun and Moon for marriages 'thinking that in nature the first marriage is the conjunction of the Moon with the Sun.'[19]

Full Moons were for culmination and fruition, the inspiration of the Muse or the terror of lunacy, also a time for weddings and giving birth. Waning Moons were for diminishing, fasting and reflecting upon meaning; Dark Moons were for ending, mourning the death of the old and imagining the form of the new. In the dark phase, when the Moon went into the underworld — either for union with the Sun, or because the Sun had murdered the Moon, or because of a wasting sickness — life waited till it came back. Journeys were postponed, even birth was postponed if possible: 'No Moon, no man' was a saying current in Cornwall in the nineteenth century. The three days of death followed by rebirth became a symbol in many of the world's myths.

It seems to make sense that all the Moon's ancient powers were derived from its *eternal* being, its capacity for rebirth. The 'water of life' — the dew, springs, rains and moisture — came from the ambrosial cup which held the nectar of immortality, which was, of course, the curved crescent of the reborn Moon. The fertilizing powers over conception and birth came from its ability to give birth to itself, as did its sway over the waxing and waning of plant life, where, most evocatively, new life appears to be born from the death of the old. As the ruler of time and the quality of that time, it inevitably became the ruler of fate: the Moirai, the three Fates in Greece, were lunar goddesses, spinning life into being, weaving its pattern, and cutting its thread.

Solarization

However, at different times around the world, a process which has been called 'solarization' set in, whereby the entire symbolic system was reversed, and the Sun became the carrier of the image of eternity while the Moon became time, now officially divorced from its eternal being.

Figure 5: Tiamat as the Serpent Mother with the Crescent Moon upon her head pursued by Marduk, holding thunderbolts and lightning. Serpentite cylinder seal, c. *ninth-eighth century* BCE.

To summarize briefly, there seem generally to have been four stages to solarization: firstly, the discovery of agriculture, which made the Sun the food-bringer as well as the Moon (though the impulse of growth was still attributed to the Moon); secondly, the subordination of matriarchal goddess-oriented cultures to patriarchal tribes, which looked to the impersonal unchanging patterns of the heavens for their model and ideal; thirdly, the development of science, which worked out that the Moon's light was reflected Sunlight; and fourthly, the advent of Christianity, where Christ was identified with the Sun and his human mother Mary with the Moon.

The key text for the conquest of the Earth and Moon goddesses and gods by the Sun and Storm god is the Iron Age Babylonian epic of creation, the *Enuma Elish* (meaning 'When On High'), which celebrates Marduk's defeat of his great-great-great-grandmother Tiamat. The myth was composed around the time of Hammurabi (*c.* 1700 BCE), not long after the Babylonians, who were non-agricultural, nomadic tribes, worshipping sun, storm and wind gods, had invaded the Sumerians, who lived in a settled agricultural, society, worshipping Moon and Earth goddesses and gods. The significance of the myth lies in the fact that it was the first creation myth to imagine the Earth as dead matter and to dramatize murder with relish, in an age, unsurprisingly, when the god's earthly counterparts were doing the same.

Here, Tiamat, the original Mother of All, is drawn as a serpent with the Crescent Moon upon her head. However, this is no longer the magnificent lunar serpent of regeneration, which sloughs its skin as the Moon its shadow, but the dragon-demon in flight who had threatened chaos — an image already resonant of the victor's point of view.

Marduk then begins again, establishing the year, making the constellations, appointing the Sun and Moon to their places in the sky, and instructing the Moon, Nanna-Sin, on his behaviour throughout the month. Having split Tiamat in two, one part to be earth and the other part to be sky, he then distributes the pieces of her lifeless body:

> He heaped up a mountain over Tiamat's head,
> pierced her eyes to form the sources of the Tigris and Euphrates ...
> Her tail he bent up into the sky to make the Milky Way,
> and her crotch he used to support the sky.[20]

The relationship of Moon to Earth, even their implicit rhythmic identity, had been severed. It was as though the Moon was now to be confined solely to the celestial sphere in the name of the new heavenly patterning, which was of a superior order to Earth.

Later, it was discovered that the Moon did not shine with its own light but with the light of the Sun. Franz Cumont in his *Astrology and Religion among the Greeks and the Romans*, points out that 'Sun-worship is essentially a *learned* cult (my italics): it grew with science itself.' It did not arise instinctively; it had to be calculated. Furthermore, this science was undertaken by the priests, not the people, and was made possible by 'continually placing [the Sun] farther and farther off in space.'[21] As a 'learned cult,' myths of the Sun belong to a later stage in the evolution of consciousness, and it is significant that the learning is only possible when the heavenly bodies are viewed from a greater distance, farther off in space, as a correlative, perhaps, to the consciousness of human beings moving farther away from their immediate instinctive experience on Earth. Only then was it possible to conceive that the light of the Moon was reflected sunlight. This was known in Egypt, and first appeared among the Greek pre-Socratics, particularly Anaxagoras, becoming commonplace in later Greek and Roman thought. Yet any new idea is slow to take hold, and it was still being discussed by Augustine in the fifth century CE.

This discovery had the radical implication that the human senses did not yield the truth. Even now it is hard to feel that moonlight does not 'come from' the Moon — that it is just 'moonshine,' in the altered meaning given to the word. Truth, it had to be inferred, must lie elsewhere, beyond the flux of time and the phenomenal world of becoming. This point of view was transmitted to western consciousness chiefly through Plato (for whom 'all change is a dying') and Aristotle, who arrived in the west in the twelfth century in Latin translations, some of them from Arabic, and finally Cicero, who wrote:

> Below the Moon there is nothing but what is mortal and doomed to decay, except the souls given to men by the bounty of the gods, whereas above the Moon all is eternal.[22]

In early times the Moon was Earth's star, a better Earth in Heaven. But when the light of the Moon was discovered to be the reflected light of the Sun, many, if not most, of the powers and stories originally attributed to the Moon were transferred to the Sun, at least as official priestly priorities, though the folk continued to live by the Moon just as they had always done. By Roman times, the Sun served as the model for earthly aspiration, carrying the image of eternity which once was held by the Moon. But when the Moon held the image of eternity for the people who lived beneath it, eternity was not opposed to life and death: it was that in which life and death inhered. Blake's evocative phrase that 'eternity is in love with the productions of time' would have been quite at home in a lunar cosmology.[23]

Many Moon stories became Sun stories, and old Moon stories were forgotten. Who now would consider the hero myth to be originally a lunar myth? It is the Sun, we are taught, who slays the demon of the dark and is victoriously reborn each dawn and each year; and it is the pattern of the Sun which the hero imitates, descending into the dark in his war of light against the dragons of darkness and death. But is it not strange that so many hero myths have their heroes dying for three days before they are reborn — imprisoned in the underworld, hung naked upon a hook (Inanna-Ishtar), swallowed in the belly of the whale (Jason, Jonah, Raven), or descending into hell (Jesus) — those same three days when the Moon is dark and was thought to be dead? It may not always matter whether the Sun or the Moon is taken as the model for the hero, but there is a significant difference of inflection: for if the essence of the hero myth is lunar rather than solar, the final emphasis of the drama falls not on conquest but on transformation. So, while the drama of heroism may appear solar to the modern solar-oriented mind — overcoming the dragon of darkness, the enemy, in a life or death struggle — its roots remain irrevocably lunar, meaning that the darkness is also a dimension of the total being which the hero has to become in order to know himself.

Evolution of consciousness

It seems that the movement in consciousness from Moon-orientation to Sun-orientation, from the late Bronze Age onwards, involved much more than a relocation of the source of fertility: it expressed, and in turn continued to foster, a change of values, which subtly but inevitably set consciousness apart from the world in which it lived and moved and had its being. The philosopher Owen Barfield describes this movement as a withdrawal of what he calls 'original participation' with nature. In his book *Saving the Appearances*, he sees the evolution of consciousness as falling broadly into three different phases, which he describes as different kinds of participation of humans with their world: 'original participation'(Jung's *participation mystique*); 'withdrawal of participation;' and 'final participation,' defined as a re-creation of the participating mind through the Imagination.[24]

In 'original participation,' the reality behind or within all the different kinds of natural life was felt to be the same reality as the human reality, so humanity and nature did not have to be apprehended by different modes of cognition. What, in contemporary terms, we would call the objective, natural world, and contrast to the subjective, human world, were once inexorably bound together, so that Nature was more awesome — both loving and terrifying — *and* more personal — peopled with divine presences imagined in human, animal and plant form. This is immanence, where the visible appearance and the invisible source are one and the same.

There followed the second stage of the pattern, beginning around 2000 BCE, when humanity began to withdraw this immanent divinity from Earth and Moon, and placed it firstly in the patterns of the heavens, with the Sun as the ruling force, and later in the invisible world, transcendent to all of Nature. Numinosity, especially in the Judaic tradition, was now found in what could *not* be seen or touched, typically expressed in Job's confession:

> If I beheld the sun when it shined, or the moon
> walking in brightness;
> And my heart hath been secretly enticed, or my
> mouth hath kissed my hand;
> This also were an iniquity to be punished by the judge:
> for I should have denied the God that is above.[25]

This numinosity also passed to the invisible world of human beings, specifically in the human capacity to internalize divinity — as humans conceived it — most obviously as the Word of God in Judaism, but also as the Rational Mind in later Greek thought, and much later as the saving power, so it was supposed, of Reason in the Age of Enlightenment. Consciousness could then expand inwards, and learn to name and order its inner and outer worlds with detachment and objectivity. Mythologically, this was the time when Sky, Storm and Sun gods replaced Earth Goddesses — Enlil and then Marduk in Mesopotamia, Atum-Re and Ptah in Egypt, Yahweh-Elohim in the Old Testament, Zeus and Apollo in Greece. Typically, the old Mother Goddess — if she was there at all — was seen as dark and chaotic, who had to be slain for the sake of light and order. The story of Adam and Eve in the Garden registers this disruption of the original bond with Nature. The solarization of the Moon falls into this stage. In many places the once male Moon god became female, and the Sun, often formerly a goddess, became male.

This process was further advanced through Christianity, with its transcendent god, whose 'only begotten son' was proclaimed by Constantine to be the *Sol Invictus* (Invincible Sun) in place of the Roman Sun god Mithras. Sinners were advised to pray to Mary who was symbolized by the Moon, which was now doctrinally fallen like the rest of nature and rested beneath her feet as a down-turned crescent, often with the black claws and tail of a dragon.

This stage of consciousness is almost too close to us to allow us to do anything but instinctively celebrate it. The question is how to reunite ourselves with the natural world without forsaking the supreme achievements of the last 2,000 years — the gradual differentiation of the human intellect, the hard-won autonomy of will and reason, wrested from the grip of superstition and religion, the painful creation of interiority and the forging of the individual in counterpole to the collective norms of the tribe. It was both a condition and a consequence of these discoveries that the objective world would lose its numinosity, and that disenchantment with nature would bring alienation and arrogance, together with a yearning to return to the original ground of being. 'Final Participation,' which hopes to restore the old participating consciousness at a new level, becomes possible when we realize that, as well as having their own reality, however that is to be conceived, Moon and Sun are, and always have been, metaphors of human consciousness.

Solar and lunar consciousness

Thomas Mann, in *Joseph and His Brothers*, coins the phrase 'lunar syntax' (or 'moon grammar') to describe how Jacob can be telling Joseph about two different men as though they were one person: 'But daylight is one thing and moonlight another ... Things look differently under the moon and under the sun, and it might be the clearness of the moon which would appeal to the spirit as the truer clarity.'[26] 'Solar' and 'lunar' have inevitably become metaphors for two alternative ways of thinking and being. For where 'solar syntax' divides, 'lunar syntax' merges, and the on-going debate between them reveals contrasting models of value, reflecting in part their different relation to darkness. Darkness is one of the modes of the Moon's divinity. But even when the Sun is given a fluid and feminine character she has to remain apart from darkness which flees before her as she comes towards it. In solar myths where the Sun is a God, the Sun is seen as heroically independent of his origins, captured, by the human psyche, forever at high noon on a cloudless day.

As the Sun sends the dark away, so the solar view sets life and thought into opposites — light *or* dark, true *or* false, good *or* evil — almost as though they were distinct entities in themselves. 'Lunar consciousness' points to the fluidity and evanescence of forms: it allows something to be and not be at the same time. Light and dark are present together in a continuing drama of expansion and contraction. Only at the poles of Full and New is there light without dark and dark without light. Symbolically, the Moon evokes the imaginative, contingent, ambiguous world of becoming — the lived-in time — in contrast to the Sun's absolutes of the ideal world of being.

Since, in western culture, the Sun is dominant and mostly male, 'solar consciousness' has long carried the culture's formal identification with its values, allying them with the virtues of clear and distinct ideas, and ultimately with 'reason' as the expression of the highest value. Mircea Eliade comments that at the end of this long process of solarization, '*sun* and *intelligence* will be assimilated to such a degree that the solar and syncretistic theologies of the end of antiquity become rationalistic philosophies; the sun is proclaimed to be the intelligence of the world.'[27] 'Lady, by yonder blessed moon I vow...,' protests Romeo, to which Juliet replies: 'Oh swear not by the moon, th'inconstant moon/ That monthly changes in her circled orb, / Lest that thy love prove likewise variable.'[28]

It seems that the perennial debate between intuition and intellect, passion and order, spontaneity and law, finds expression in these metaphors of Moon and Sun. Where the Moon offers a model of completeness, the Sun offers a model of perfection. Accordingly, the Moon has often been drawn as an image of soul, with the Sun as an image of spirit, though these metaphors have been complicated by the Judaeo-Christian priority of according a higher value to spirit than to soul.

Figure 6: Mercurius reconciling the opposites in the Caduceus. Chymisches Rosegarten, Virdarium Chymicum, *Frankfurt, 1624 (Mannheim University Library).*

Nonetheless, *both* modes of consciousness are necessary for human wholeness, and in the myths and rituals of the world there is everywhere an impulse to bring the Moon and Sun together as equals: in earlier times through the 'Sacred Marriage'— *Hieros Gamos* — of Moon and Sun (whichever heavenly body held the image of eternity); and later through the symbolic life which seeks to reconcile lunar and solar ways of being. The point of emphasizing the lunar mode here has been to attempt to restore the balance. Just as stories of the Goddess have to be reclaimed from stories of the God, so myths of the Moon may now appear in their own light not only told from a solar point of view.

In Alchemy, the aim was to work with both lunar and solar ways of knowing and being, ultimately uniting them in a third state which the Alchemists called 'Stellar Consciousness,' an integrated existence that the Egyptians called 'Living in Truth,' and we might call the Imagination.

Notes

1 Yeats, W. B., 'He Wishes for the Cloths of Heaven,' *Collected Poems* (London: Macmillan, 1952), p. 81.
2 Shakespeare, William, *The Merchant of Venice*, V.i.54–65.
3 Coleridge, S. T., in R. Holmes, *Coleridge: Darker Reflections* (London: Harper Collins, 1998), pp. 38–9.
4 Campbell, Joseph, *The Hero with a Thousand Faces* (Bollingen Series XVII, Princeton: Princeton University Press, 1968), p. 13.
5 Jung, C. G., *Collected Works*, 20 vols., Sir Herbert Read, Gerhard Adler, Michael Fordham, William McGuire, (eds.), R. F. C. Hull, trans. (London: Routledge & Kegan Paul, 1957–79), 9:1, p. 6.
6 Origen, *Homilae in Leviticum*, in Jung, *CW* 16, p. 197.
7 Shakespeare, *As You Like It*, II.vii.165–6.
8 Quoted in Sachs, Miriam, *The Moon* (New York, London, Paris: Abbeville Press Publishers, 1998), p. 34.
9 Plato, *The Timaeus*, 37.C.D.
10 Marshack, Alexander, *The Roots of Civilization* (London: Weidenfeld & Nicholson, 1972).
11 Briffault, Henri, *The Mothers* (London and New York: George Allen & Unwin Ltd., 1927), iii, pp. 3–4.
12 Budge, E. A. Wallis, *The Gods of the Egyptians* (New York: Dover Publications, 1969), i, p. 412.
13 Frankfort, Henri, *Kingship and the Gods* (University of Chicago Press, 1948), p. 196.
14 Jacobson, Thorkild, *The Treasures of Darkness: a History of Mesopotamian Religion* (New Haven, Connecticut: Yale University Press, 1976), p. 122.
15 Jacobson, pp. 138–9.
16 Hollis, A. C., *The Nandi: Their Language and Folk-lore* (Oxford: 1909), pp. 52–71.
17 References for this section in Cashford, Jules, *The Moon: Myth and Image* (London: Cassell Illustrated, 2003), pp. 38–67.
18 Jung, *Modern Man in Search of a Soul* (London: Routledge & Kegan Paul, 1981), pp. 129–30.
19 Plutarch, *Moralia*, XV.105.
20 Jacobson, p. 179.
21 Cumont, Franz, *Astrology and Religion among the Greeks and the Romans*, J. B. Baker, trans. (New York: Dover Publications Inc., 1960), p. 71.
22 Cicero, Marcus Tullius, *De Republica de Legibus*, T. E. Page, *et al.*, eds., Clinton Walker Keyes, trans. (Loeb Classical Library, Cambridge, Mass., and London, Harvard University Press and William Heinemann Ltd., 1952), VI.17.17.
23 Blake, William, *Complete Poetry and Prose*, Geoffrey Keynes, (ed.) (London: Nonesuch Press, 1961), p. 183.
24 Barfield, Owen, *Saving the Appearances: A Study in Idolatry* (Hanover, New Hampshire: Wesleyan University Press, 1989), passim.

25 *Job*, 31:26–8.
26 Mann, Thomas, *Joseph and His Brothers* (London: Secker & Warburg, 1956), p. 77.
27 Eliade, Mircea, *The Sacred and the Profane*, Willard R. Trask, trans. (New York and London: Harcourt, Brace Jovanovich, 1959), p. 131.
28 Shakespeare, *Romeo and Juliet*, II.I.151–3.

6. The Soul of the Sky

NOEL COBB

This paper will explore the difference between living with a sense of a sky ensouled and living with a secularized world view of sky.

> Today I announce a new plan to extend a human presence across the solar system ... It is time for America to take the next steps. Mankind is drawn to the heavens for the same reason we were once drawn into unknown lands and across the open sea. We choose to explore space because doing so improves our lives and lifts our national spirit.
>
> G. W. Bush[1]

> We cannot fully exploit space until we can control it.
>
> D. Rumsfeld[2]

> **Juliet**: It was the nightingale, and not the lark,
> That pierced the fearful hollow of thine ear;
> Nightly she sings on you pomegranate tree:
> Believe me, love, it was the nightingale.
>
> **Romeo**: It was the lark, the herald of the morn,
> No nightingale: look, love, what envious streaks
>
> Do lace the severing clouds of yonder east:
> Night's candles are burnt out, and jocund day
> Stands tiptoe on the misty mountain tops.
> I must be gone and live, or stay and die.
>
> W. Shakespeare, *Romeo and Juliet*[3]

What *is* the sky? When does sky become space? Its everyday ever-presence could lead us to take it for granted and be, like the air, *simply there.* But the title of this conference prompts us to reflect on the sky's actual and essential nature, and the more we do so, the more mysterious it appears.

It is there during the night when we sleep; it is there in the day when we wake. We learn to know it in all weathers and seasons, and even if we are imprisoned in a lightless cave, the memory of its existence consoles, inspires or fills us with emotions as contrary as longing and dread. Goethe, it is said, spoke of holding one virtue before all others: *Reverence.* Reverence for what is above us, what is around us and what is beneath us. Apart from the sky, if we were to give a name to 'what is above us, around us and beneath us' as well as in us, we could do no better than to call it the Soul of the World, or, to give it its ancient Greek name, *Psyche tou kosmou.* Ancient alchemical wisdom says the greater part of the soul lies outside the body. Heraclitus proclaimed: 'No matter how far you may travel in any direction, you will never come to the edge of the soul.'[4]

In the *Enneads*, Plotinus quotes Sophocles' blind Oedipus at Colonus: 'All the place is holy,' adding, 'and there is nothing without a share of soul.' Plotinus is unambiguous on this: 'For there never was a time when this universe did not have a soul, or when body existed in the absence of soul, or when matter was not set in order.' In another image he says:

> It is as if a net immersed in the waters was alive, but unable to make its own that in which it is. The sea is already spread out and the net spreads with it, as far as it can; for no one of its parts can be anywhere else than where it lies ... wherever body extends there soul is.[5]

Making absolutely sure we do not simply understand the intelligible world — which informs the soul — as being outside and above the world of the senses, Plotinus says that although 'the sense-world is in one place, the intelligible world (the *Nous)* is everywhere.'[6]

Now, Earth and Sky comprise a primordial, dyadic unity: something traditionally understood in most ancient cosmologies, where the two are personified as male and female, a couple. In any case, seeing Earth and Sky as *ensouled* means, first of all, apprending their essential reality as *image.* Second, it means calling for that activity specific to soul: *personifying* or 'imagining things.' This is basic to all questions of soul.

Robert Sardello speaks of 'imaginal worlds, populated by the composing beings of the fabric of the physical planet,' saying, 'our imagination is the organ by which we know these composing beings … the psychic world is this world of the physical world and none other.' There is a:

> feeling we all have but never acknowledge, that everything is animated. Such feeling is not animism, for animism is a theory that says that soul life is projected onto an inanimate world from within the human psyche. On the contrary, soul inheres within the world and creates our psyches.[7]

For Martin Heidegger, Earth and Sky are part of a primal oneness that is divided into four. Humans and divinities complete that fourfold oneness. *Human being,* then, consists in *dwelling* and, indeed, dwelling in the sense of *the stay of mortals on the earth.* But *on the Earth* already means *under the sky*. Both of these also mean *remaining before the divinities* and include a *belonging to men's being with one another.* By a primal oneness, the four — Earth and sky, divinities and mortals — belong together in one. 'Earth is the serving bearer, blossoming and fruiting, spreading out in rock and water, rising up into plant and animal.' Heidegger continues:

> The sky is the vaulting path of the Sun, the course of the changing Moon, the wandering glitter of the stars, the year's seasons and their changes, the light and dusk of day; the gloom and glow of night, the clemency and inclemency of the weather, the drifting clouds and blue depths of the ether. When we say sky, we are already thinking of the other three along with it, but we give no thought to the simple oneness of the four.[8]

The sky, like the rest, then, is ensouled. The soul of the sky is not an element of the physical sky, to be found by chemical analyses of its gases, molecular studies of stardust or deepspace telescope searches for evidence of black holes or the Big Bang. Sky — our sky — is, since Aristotle, not soul-less, placeless space, an emptiness, but place, and place as 'the first of all things' which 'has a power of its own to change the course of events.'[9] To see the sky as space is to empty it of place and thereby soul.

First of all, the soul of the sky is to be found in its *innerness.* This innerness is not something *in* the sky at all, not in its atomic structures or in the astronomical measurements of its immensely complex, inter-related movements as if it were one vast clock that we could finally explain. The sky's innerness lies more in the natural poetry of its self-display and in what it says to us mortals who are under it, with one another, during our brief stay on earth. Take rain — rain is one thing for the scientist, but another for the indigenous rainmaker. Mohammed said that every raindrop that falls is accompanied by an Angel, for even a raindrop is a manifestation of Being. For ages, traditional wisdom has called rain 'the elixir of life.'

If the soul of the sky is given by images, 'those images that yet, fresh images beget,' then we need not call a halt to the soul's innate capacity to imagine itself.[10] In the boundlessness of soul there is room for everything. Heaven, says Plotinus, gains its value by the indwelling of soul. 'As the rays of the Sun light up a dark cloud, and make it shine and give it a golden look, so soul entering into the body of heaven gives it life and gives it immortality and wakes what lies inert.'[11] Many artists, like Constable, can arrive at only one conclusion: the sky is the source of light in nature and governs everything.

The beautiful image Plotinus gives of the All, as a 'single living being which encompasses all the living beings that are within it,' does not exclude the existence of other images. It is not trying to prove or explain anything, but to better imagine things as they are.[12] 'The first idea was not to shape the clouds in imitation./ The clouds preceded us/ There was a muddy centre before we breathed./ There was a myth before the myth began,/ Venerable and articulate and complete./ From this the poem springs.'[13]

For Plotinus, 'This one universe is all bound together in shared experience and is like one living creature, and that which is far is really near ... and nothing is so distant in space that it is not close enough to the nature of the one living thing to share experience.'[14] 'This All,' at the same time 'is visibly not only one living creature, but many.' He compares the 'figures of the heavenly circuit,' or constellations, to individual dancers, each moving in harmony with the whole 'as if they were performing a single ballet in a rich variety of dance-movements' — Plato's 'choreography of the Gods.'[15] 'In the All,' Plotinus says, 'there is an indescribably wonderful variety of powers, especially in the bodies which move through the heavens.'[16] There is power both in the figures, or constellations, themselves, as well as in the heavenly bodies, or stars, which compose them; as with dancers in whom each hand and limb displays distinctive powers as well as there being great power in the dancer herself.

For Plotinus the truest and most satisfying image for the universe as a whole is an *aesthetic* one. The different movements of the Dancers of the All 'serve the dance and help to make it perfect and complete.'[17] Marsilio Ficino, who clearly based his famous study 'On Obtaining Life from the Heavens'[18] on an inspired reading of this *Ennead*,[19] shares this aesthetic sense of the cosmos. As Eugenio Garin said, 'The world as a work of art could be the title of all Ficino's philosophy.'[20]

Ficino claims that to receive the most from the very spirit of the life of the world, through the rays of the stars, our spirit must be properly prepared and purged.[21] This *spiritus* is innate in everything from plants and trees, seashells and stones, minerals, metals and the elemental energies of the celestial bodies. It diffuses its rays, which are not only visible but can also see through the stars which act as its eyes. The notion of the *stars as eyes* is a soul-image. To see the stars as eyes, we need to stop the habit of seeing as a subjective activity. It must give way to another kind of perception — no longer looking at a thing but being seen by it. The stars as watchers, witnesses. 'When things look at us, they reveal, manifest themselves to the heart.'[22]

In another essay, Ficino reverses the idea of the 'light of reason' to become 'the reason of light;'[23] light comes first, then 'our reason.' The reason of light is a gift of the divine, not a product of our making. The quality of this kind of reason is its connection to the deep psyche: the intelligence of 'innerness,' something far deeper than ego consciousness.[24] This intelligence, for Ficino, is *risus coeli,* 'the Laughter of the Sky,' and he imagines the Sun as the eye of the heavens, an eye that conveys true spirit, as does the eye of the lover ... or eyes in a smiling face. 'Consider the difference,' Thomas Moore suggests:

> between a communication received from another person in words and the message conveyed through facial expression or even a twinkle in the eye. This latter is the analogy for the kind of spiritual insight Ficino would have us pick up in the world itself — an awareness of the twinkle in the eye of the cosmos.[25]

In the Homeric *Hymn to Demeter,* Earth grows a wonderfully bright narcissus to trick Persephone as a favour for Hades. The poet says that this flower 'astonished everyone who saw it, mortals and immortals alike. It pushed up a hundred heads, and the fragrance of it made the vast sky above, and all the earth and the sea, laugh.' A suspicion of collusion between world and underworld gives the classic abduction story a twist, and the laughter of sea, Earth and sky seems cruel when we consider Persephone's fate. Compare Plotinus: 'The life of the universe does not serve the purposes of each individual but of the whole' and 'does not always give each individual what it wants.'[26]

Heidegger reminds us that 'The nature of the image is to let something be seen ... Poetic images ... are visible inclusions of the alien in the sight of the familiar. The poetic saying of images gathers the brightness and sound of the heavenly appearances into one with the darkness and silence of what is alien.'[27]

'While psychology still locates imagination within the recesses of subjectivity, the word *imagination* belongs to the side of the world,' says Sardello.

> The Indo-European root of the word imagination refers to anything changing or intermittent, capable of catching or fixing one's attention. The Sanskrit root of the word refers to the ever-changing, ensnaring play of appearances. Imagination also means something fleeting, like a cloud, or that which winks or signals. To approach the multiplicity of the things of the world requires imagination because they are imagination in action — things changeful, spellbinding, shuddering, arresting, magical, tricky, elusive, shimmering.[28]

The Egyptians personified the changeful nature of the sky — the arresting, magical, elusive, shimmering imagination of the world-soul — in the image of Nut, the sky-goddess, seeing her as arched naked over the earth, facing down with hands and feet touching the eastern and western horizons. Sometimes represented as a suckling sow, she was mistress of heavenly bodies. Her children, the twinkling stars, entered her mouth and emerged from her womb, giving an animal sense to the sky as 'the female pig who eats her piglets.' As mother of the Sun-god, Re, Nut also swallowed him each evening and gave birth to him the next morning. Nut was often engraved or painted on the underside of a coffin lid, as if to say the deceased was now under her protection and, like the Sun, would one day be reborn after a passage through her Kingdom of Heaven.

The theophany of Isis, Queen of Heaven, is pictured by Apuleius as the figure of a woman rising from the sea, garlanded with blossoms, a crown of rearing snakes and waving ears of corn.[29] The three colours of her dress are radiant-white, saffron-yellow and rosy-red. Leaving her right shoulder uncovered — like images of the Buddha — a night-black cloak envelopes her. Stars glitter over its surface, and a full Moon blazes among them. She wears sandals of palm leaves and her breath has all the fragrances of Arabia. Could we find a more fitting way to personify the soul of the world than this image of Isis?

These images speak of the life and soul of the cosmos around us.[30] They are grand yet, at the same time, intimate and near. Humans have prayed to them for ages. Are they merely remnants of mankind's superstitious childhood, to be let go of now that we are the 'rulers' of the universe? Heidegger, deepening his thoughts on dwelling, says: 'Mortals dwell in that they receive the sky as sky. They leave to the Sun and the Moon their journey, to the stars their courses, to the seasons, their blessings and their inclemency; they do not turn night into day nor day into a harassed unrest.'[31]

In *The Tempest*, Shakespeare expresses this humility in the words of Gonzalo, rebuking the courtiers who mock his care of the soul: 'You are gentlemen of brave mettle; you would lift the Moon out of her sphere, if she would continue in it five weeks without changing.'[32] Or we could imagine the poet, Gerard de Nerval, saying to George Bush, about his deadly, plutonium-fuelled, 'Prometheus Project'[33] for the exploitation and colonization of space:

> Free thinker! Do you think you are the only thinker/ on this Earth in which life blazes inside all things?/ Your liberty does what it wishes with the powers it controls, but when you gather to plan, the universe is not there.[34]

NASA's loveless attitude for the Earth is clearly mirrored in our unconsciously accepted belief in heliocentrism. Yet, 'the famous physicist, Ernst Mach, once contended that there are no physical grounds on which to argue that the Earth is rotating around the Sun, since the relativity of motion equally well allows us to say that the universe is rotating around the earth.'[35]

A thoughtful perspective on heliocentrism is given by Ginette Paris, who says:

> The cult of Hestia is linked to the geocentrism of the Greeks. If the archetype of the house, signifying a return to the center, is one of the important archetypes of our *psychological* life, it is understandable that the idea of our

> planet as the center of the universe is just as important in elaborating the *collective* values associated with 'our' planet. We could hypothesize that in losing geocentrism we have lost the feeling that this planet is our home; we have lost a bond with her ... We see two ideas reappearing simultaneously: the first is that the Earth is not the center of the universe. The second is that of a departure: one day we shall leave this little planet and explore all other possible ones.[36]

The consequences of our drive to go further up and out has allowed our planet, our collective household, to deteriorate. 'It now takes ecology, "domestic science," to remind us to take care of our planet, as if once she ceased to be the center of our attention she became a peripheral fact, something we could leave after using it.'[37] For Hestian, geocentric consciousness, 'our' sky is the sky of 'our' Earth. 'Our' sky is not a sky with six purple moons, nor a sky where fanged mushrooms fly through deadly methane outside a dome of controlled light and weather: 'Our' sky enfolds 'our' home, 'our' Earth. The first house — said to be built by Hestia — echoes the vault of the sky in its roundness, a roundness perennially honoured by the great Goddess-Domes: The Hagia Sophia, St Mark's of Venice, St. Paul's of London, the Duomo of Florence, the blue mosque of Istanbul and the Taj Mahal of Agra. Our sky is *finite* as well as infinite. At night it recalls Plotinus' great dancer, but the skies under which we live our little-life-of-everyday is rounded with a sleep and is also a particular life here-with-each-other. These skies are a deeply integral part of our griefs and joys. Our sky is above all an emotional sky.

Think of twilight, that time *inter canem et lupem* — between the dog and the wolf. Have you never felt the prickling at the back of your neck at its onset? Its luring yet uncanny ambiguity? By turns sensuous, playful, expectant, dangerous, deceitful, or ghostly, oracular, protean and indefinable, twilight is a crack between the worlds; a mysterious, fluid time-boundary: not this, not that; an archetypal arena between the demands of the day and the enigmas of night; a dimming of blinding solar consciousness. Can you ever forget that moment at twilight in Bodhgaya — seat of Sakyamuni's awakening — when a huge red sun sank below one horizon, while a round ivory moon rose, opposite?

And what is it about dawn? You sit cross-legged at the side of an inlet where the waves scarcely move. Light from the Moon has brushed a path of liquid silver across the bay, as she slowly passes overhead to step behind the cliffs. Out of the darkness, hours later, a first gleam of light gives a faint blush to the sky in the east. The light brightens now, as if heralding the arrival of a king. The Earth holds its breath, awaiting his appearance. Then, on the lip of the horizon, the glimmer becomes the sheerest sliver of burnished gold. Like the head of a baby crowning its mother's vulva as it emerges from her womb, the Sun pushes free and out of the night. Dawn brings with it such human joy, such fresh hope. As Sappho says: 'Standing by my bed/ In gold sandals./Dawn that very/ Moment awoke me.'[38] And yet, dawn can, as for Romeo, also bring danger.

The Moon also rises, but her light is colder, more hypnotic. Not so much human joy as a sense of ecstatic elation. The night air is chill and still. Father holds you in his arms, wrapped in soft wool. He has taken you out of the light of the house into the dark outside, and now he points overhead. Your eyes follow. There is nothing there but that bright, glowing ball. All around, the chill night waits. Does it wait? Father's breathing is the only sound. Then he says in a hushed voice: 'Look! Look at the Moon!'

The miracle of imagination works its magic. Why does the Moon play such a part in the psyche? It plays such a part because it reminds us of change. Empty to full. Full to empty. This change is different from the change of night to day. This change, like the Sun's, is about cycles of death and rebirth; as in breathing, but slower, deeper as with the tides the Moon creates through its great attraction; as in lunar beings like women and the cat:

> The cat went here and there
> And the moon spun round like a top
> And the nearest kin of the moon,
> The creeping cat, looked up.
> Black Minnaloushe stared at the moon,
> For wander and wail as he would,
> The pure cold light in the sky
> Troubled his animal blood.
> Minnaloushe runs in the grass
> Lifting his delicate feet.
> Do you dance, Minnaloushe, do you dance? ...
> Does Minnaloushe know that his pupils
> Will pass from change to change,
> And that from round to crescent,

> From crescent to round they range?
> Minnaloushe creeps through the grass
> Alone, important and wise,
> And lifts to the changing moon
> His changing eyes.[39]

When the Moon is away, the night sky is darker than indigo, and the desert air is crystal. I lie awake on one elbow in the Saharan sand-sea called The Grand Erg, beside me a star map and a torch. I've turned the torch off and lain back, whispering the names of the stars I've learned to locate among the masses above. The sight is overwelming. Here, infinity has an image, it isn't just mathematics. The spirit may long to calculate how many millions of stars there are, but the soul delights in their self-display. For me, it all started with Merak and Dubhe in The Great Bear, the first two my Dad showed me. Later came the North Star, Polaris. Then, under the Mediterranean sky, Cygnus, the swan, flying with wings wide and neck outstretched through the Milky Way. Further South, I found Orion with his sword and belt, resplendent with those incandescent eyes: Betelgeuze, Bellatrix, Rigel and Saiph. What wonder! And, all hail! Sirius, Dog Star, brightest star in the sky. And the constellations, Castor and Pollux. Taurus the Bull, with its giant orange Eye, Aldebaran; Hercules, forever running the stellar Olympics; and the Pleiades, those lovely, veiled daughters of Atlas. Bootes, the herdsman, where the great daimon, Arcturus, lives. And, high in the spring night, Leo, guarding Regulus between his paws; five times bigger, they say, than the Sun and 160 times as bright.

For how many centuries have we looked up at the stars? The constellation of Leo, lion *couchant,* seems ageless. Who saw it first? And where? In the deserts of Sumer? On the rooftops of Babylon? At sea in the Indian Ocean? The *Oxford English Dictionary* states flatly: '*Constellation:* a number of fixed stars grouped together within the outline of an imaginary figure traced on the face of the sky.' Yet, it lets slip a prime indication of *anima mundi*: the Face of the sky. Question: When we study the face of a friend do we see imaginary lines traced on her face — or do we read the character of her soul?

Perhaps, as Plotinus taught, in this one universe, all bound together in shared experience, that which is far is really near. That night so many years ago I hadn't read Plotinus. Nor had I yet heard Sir Laurens van der Post say the Kalahari bushmen felt 'known by the stars.' But in that great desert night, the sense that our planet is *known* by the stars was suddenly very real. Our heliocentric culture had taught me our Earth was 'insignificant.' I couldn't really believe this. Yet, who were we? And could we hold up to the stars something worthy of their attention? The question seemed unanswerable. Then the thought came: music! I had a small tape recorder on which, lonely for familiar sounds, I would sometimes play recordings. But since arriving in the desert, mortal compositions seemed out of place. I'd tried many favourites: classical/modern. Nothing could stand before that sky. At last, I remembered a recording of Verdi's *Requiem,* conducted by Barbaroli. These sounds, these voices, might be beautiful enough to reach the heart of a listening star. If 'Mozart is sunshine,' as Dvorak said, then we must add that 'Verdi's Requiem is starlight!' Yes, this music was worthy of the stars!

Daily climbing higher in the Nepalese Himalayas, we had reached the last outpost of civilization — Tangboche Monastery. My companion and I paused to regain our strength and adjust to the thin air. On a grassy slope surrounded by blossoming rhododendrons, we rested in contemplation. Meditation texts speak of one's 'innate Buddha nature' as the *clear, open sky.* That day, the sky was the deepest ultramarine I had ever known. Mt Ama Dablan soared above us, nearly as tall as Everest, with her necklace of dazzling glaciers gleaming in the sunshine. Gazing into the depths of this azure vault, my eyes gradually made out tiny golden sparks which formed into shapes. At first, I couldn't believe it. Then, it became impossible to doubt! The entire sky: a lattice-work of tiny, golden buddhas! In wonder, I sat with this blessed vision until, distracted, I lost my concentration. When I looked again, I saw only the most intense blue of the sky — except now the thinnest wisp of vapour was caressing the mountain's cheek. As I watched, the white wisps thickened, became cloud. I was watching the actual formation of cloud out of clear sky. Ancient meditation texts say: this is the way thoughts form in the mind. We can watch them, but we do not create them. The clear blue sky is like our 'original mind.'

The circumscribed fears and passions of mortals shrink in humble lowliness beside the sky's, and yet we strain to reach its dignity and grandeur. If the sky's emotions show in its moods, then any sky can never be one ideal, pure, stainless blue. Not for long. Its moods are ever-changing. To yearn for permanent blue skies, where all is spiritually pure and heavenly blissful, is to dream of a sky that is not of this Earth. It is to fall prey to a monotheistic fallacy: a life-negating fantasy of ultimate paradise.

Seated on a meditation platform built by hermits in the Saharan Hoggar, on the edge of a steep precipice, I became aware of the changing light on the mountains in the distance. By late afternoon, what seemed to have been black, opaque, rock-scapes began glowing imperceptibly with pigment: long rays of liquid Sun brushed dark ravines with powders of all the spectrum: rust tinged with verdigris, saffron, scarlet and vermillion, as if a rainbow had melted into the Earth. The air was motionless; nothing breathed. Miles away a Touareg and his camel slowly merged into deep mountain shadow. The whole day had been a reverie on that third century text, *Celestial Hierarchies,* the distinctions between 'orders' of invisibles. Without warning, a loud crack of wind exploded behind me, nearly toppling me over the edge. The fear I felt then was not of falling, but of experiencing Wind at its Source, the entry of spirit into the world. *Pnevma.* The Breath of the Sky. The sky speaks with many voices. One bright August morning in Norway's Rondane range, the sky grew black and utterly quiet, before one deafening thunderclap — and that was all. The Greeks would have said Zeus had spoken.

A different experience was that of the skies over the ancient kingdom of Swat where no rain had fallen for six months. On midsummer night, far to the west over the Hindu Kush, growing rumblings warned of stormclouds moving closer. For hours, lightning played like the charged crackling of a dragon's transparent nervous system as it twisted and coiled over the hills. The parched Earth waited. If rains came, the parched Earth would be charged with the glory and grandeur of Sky, the maize could be planted and the harvest would be assured. At midnight, standing on the flat roof of my Gujur house, I felt the Dragon's furious presence overhead — the sky-rolling roars and flashes of his eyes, and — with this — the heavens opened.

From the Makabeng in northern Transvaal to the New Mexican deserts of the Hopi, from the Taoists of old China to the last San bushmen of the Kalahari, rain has been drawn down to Earth by chants, cries, dances and drums: ritual prayers made in the hope of quenching Earth's thirst with the life-giving water in the sky. Who doesn't weep when the 'stars throw down their spears/ and water heaven with their tears?'[40] What better metaphor for blessings? 'The quality of mercy is not strain'd,/ It droppeth as the gentle rain from heaven/ Upon the place beneath.'[41]

How different the uncanny panic that led to Edvard Munch painting *The Scream.* 'I stopped,' he said, 'and leaned against the railing, half-dead with fatigue. Over the grey-blue fjord, the clouds hung, red as blood and tongues of flame. My friends drew away. Alone and trembling with fear, I experienced Nature's great scream'[42] And let us not forget Vincent's skies: from ominously oppressive to watchful and shining with stars; ecstatic blue spirals over twisted cypresses; hot and blazing, or doom-laden and threatening, filled with flapping flocks of crows.

The sky can deepen our dread and mirror premonitions of disaster: fine flakes of ash dropping out of brown, overcast skies telling of large-scale forest fires; sinister sky-silhouettes over burning London; clichéd pictures of the first atomic mushroom filling the sky; American helicopters napalming Vietnamese straw huts from the sky. Closer to home, RAF jets ripping through the sky overhead as we walk among the newborn lambs. In the last century, humanity has learned to fear the sky in altogether new ways: death-gas sprayed from the sky over families of fleeing Kurds, as thoughtlessly as over cornfields and apple orchards. Pests? Get rid of them! Untold millions of birds — those ecstatic celebrants of flight and song — extinguished! Whole species gone forever — because men believe they can dispense death without consequences — stubbornly ignorant of how all things are connected.

Is there a more glorious embodiment of the soul of the sky? Those 'free-flying/ leaves,/ champions of the air,/ petals of smoke,/ free,/ happy/ flyers and songsters,/ ... navigators of the wind ...' as Pablo Neruda calls them. Birds, whose tiny, feathered skulls contain, it seems, entire maps of constellations to guide them on their migrating way over seas and continents at night. As Blake says, 'How do you know but ev'ry Bird that cuts the airy way./ Is an immense world of delight, clos'd by your senses five?'[43]

And further, do birds not respond to our emotions? What, you've never heard birds join in with an opera broadcast or imitate human noises? Songbirds are original purveyors of melody, the soul's evocation of inspiration and mood. Surely nothing in the world is more expressive of joy than the pure music of the nightingale's bright song, that kind of 'brilliant calling and interweaving of glittering exclamation such as must have been heard on the first day of creation, when the angels suddenly found themselves created, and shouting aloud before they knew it.'[44] And who has not felt an ancient melancholy tug at his breast with the mournful honking of geese, passing by overhead on their long-distance wings?

Who notices the soul of the sky? In times past it was the sailor, the farmer, the star-gazer and the auger. Speaking of the Etruscan augers, D. H. Lawrence mused:

> To them, hot-blooded birds flew through the living universe as feelings and premonitions fly through the breast of a man, or as thoughts fly through the mind ... And since all things corresponded in the ancient world, and man's bosom mirrored itself in the bosom of the sky ... the birds were flying to a portentous goal, in the man's breast as he watched, as well as flying their own way in the bosom of the sky. If the auger could see the birds flying in his heart, then he would know which way destiny too was flying for him.[45]

Imprisoned by skyscrapers and filthy skies, who even sees the constellations or the morning star, the Moon or the sunrises and sunsets? Yes, we can befoul the sky with our unconsciousness. While the Victorian poor of London coughed out their lungs in smog, even great artists like Whistler and Monet, who saw beauty in those polluted skies while rooming at the Savoy, never protested outside Parliament against the outpouring of two hundred thousand tons of soot into those same skies each year. The sky dwarfs our most extreme despairs with its natural rages — hurricanes and tornadoes — demolishing all that lays before them. Yet, is there any greater measure of human grief than Lear, at the death of Cordelia? 'Howl, howl, howl, howl! O, you are men of stones:/ Had I your tongues and eyes, I'd use them so/ That heaven's vault should crack. She's gone for ever!'[46]

I have been talking of the soul of the sky, a face of *anima mundi,* and how we apprehend it. I spoke of the sky's 'innerness' in images of sky; of the soul's innate capacity for personifying itself; of learning to feel perceived by the sky; to sense *its* emotions and how intimately our emotions — whether or not we are aware of it — continually respond and correspond to it. I propose we place 'the emotional sky' at the heart of this collective gathering. This is no sentimental projection. It is an essential countermeasure to the Puer-Prometheus-Mars-Apollo drive to soar away from 'our' world and 'our' sky — into abstract space, 'somewhere over the rainbow' — whether literally or in the deadly glamorization of science and the hubris of its belief in a scientifically objective world.

James Hillman indicated that emotion is a two-way bridge between us and the world, that 'our own emotions ... are continually interacting with the world of nature,' and that;

> every 'cut' between subject and object is arbitrary. Man and world are fundamentally inseparable because the same

> energy constitutes man and world. This is evident on the physical level, that is, we are composed of the same elements as the world about us. But on the psychological level we experience the cut because we neglect our emotions and thus miss the emotional aspect of things. We have such little and such primitive emotional awareness that the world about us is either dead and without value or falsely overcharged ... We have left undeveloped the faculty of apprehending the world as emotional.[47]

John Keats feared that a scientific gaze would 'unweave a rainbow' and 'empty the haunted air' by 'conquering all mysteries by line and rule.'[48] The rainbow is the only phenomenon in nature depending for its appearance not only on the Sun but on us, moving with the observer. Thus, says Sardello, we are face to face with, or directly perceiving, the soul of the world.[49] The soul of the world, in other words, is not perceived unless we give it emotional attention, and turning that attention away from it to 'line and rule' is one way of destroying the bridge which unites above and below and whose magical outcome is the world's enchantment.

The sky — our sky — with its 'greater and lesser lights,' its moods, its stars and planets, its mists and clouds and rainbows, is our best teacher and image of Awakening. In Sanskrit, 'the one who came awake,' *fully awake,* was known as Buddha, who described himself as 'awakening to the delusion of all independently-existing things.' At the end of the *Diamond Sutra* we read: 'Thus shall you think of all this fleeting world: a star at dawn, a bubble in a stream; a flash of lightning in a summer cloud. A flickering lamp, a phantom and a dream.'

But there is something else those who have gazed long and lovingly at 'the world as a work of art' say; that whatever else it may be, the world is beautiful. To see this is to love it. So, although Shakespeare and the Buddha teach that we are such stuff as dreams are made of and even the great globe itself and all that it inherits shall dissolve into thin air, they may not have been altogether against the kind of attachment to the beauty of the world that comes from love and not from acquisitiveness and the wish to exploit that breeds envy, hatred and strife. The sky teaches something else, then, besides awakening. Love! — 'the love,' as Dante says, 'which moves the Sun and the other stars.'[50] And Jalal'uddin, our Mevlana, says: 'The soul within the soul lives in a lover./ Consider this metaphor: how you love is/ the open sky.'

Notes

1 Bush, George W., in a speech to US Space Agency NASA headquarters in Washington DC (January 2004).
2 2001 Commission to assess US National Security, chaired by Donald Rumsfield.
3 Shakespeare, *Romeo and Juliet*, III.v.1–11.
4 Heraclitus, Fragment 42.
5 Plotinus, *Enneads*, 3 Volumes, A. H. Armstrong, trans. (London: Harvard University Press, 1929), IV.iii.9.
6 Ibid, V.ix.[5]13.
7 Sardello Robert, *Facing the World with Soul* (New York: Lindisfarne Press, 1992), p. 22.
8 Heidegger, Martin, *Poetry, Language and Thought* (New York: Harper and Row, 1971), p. 87.
9 Casey, Edward, *Spirit and Soul* (Texas: Spring Publications, 1991), p. 293.
10 Yeats, W. B., 'Byzantium,' in *Collected Poems* (New York: Macmillan, 1951), p. 243.
11 Plotinus, *The Enneads,* V.i.2.
12 'The All,' in Plotinian terms, is equivalent to everything existing in the universe after the One. The quotes in this section are from Plotinus, *Enneads,* IV.iv.32.
13 Stevens, Wallace, 'Notes toward a Supreme Fiction,' in *Collected Poems of Wallace Stevens.*
14 Plotinus, *Enneads,* ibid. The whole quote is as follows: 'This All is visibly not only one living creature, but many; so that in so far as it is one, each individual part is preserved by the whole, but in so far as it is many, when the many encounter each other they often injure each other because they are different; and one injures another to supply its own need, and even makes a meal of another which is at the same time related to and different from it; and each one, naturally striving to do the best for itself, takes to itself that part of the other which is akin to it, and makes away with all that is alien to itself because of its self love.'
15 *Ibid.*
16 *Ibid,* IV.iv.36.
17 *Ibid,* IV.iv.33.
18 Ficino, Marsilio, 'De Vita Coelitus Comparanda,' in *Liber de Vita*, Book Three.
19 Plotinus, *Op cit.*
20 Garin, Eugenio, *Astrology in the Renaissance* (London: Arkana, 1990), p. 76.
21 Ficino, *Liber De Vita*, Book Three, chapter eleven.
22 Sardello, Robert, *Facing the World with Soul* (New York: Lindisfarne Press, 1992), p. 127.
23 *De Lumine.*
24 'The qualities of this light of the heavens, are: the fruitfulness of life, the perspicacity of our senses, the certitude of our intelligence, and the bountifulness of grace,' in Moore, Thomas, *The Planets Within* (Lindisfarne Books, 1990), p. 91.

25 *Ibid*, p. 92.
26 For these ideas see *Ennead,* IV.iv.38–39. Plotinus says the difficulties about 'the gift of evils coming from the gods' would be solved by considering that it is not their deliberate (single) choices which are evil, but the resultant mixture of actions as a consequence of the life of the one universe.
27 Heidegger, '... Poetically Man Dwells ...,' in *Poetry, Language and Thought* (New York, Harper & Row, 1971).
28 Sardello, *Facing the World with Soul* (New York: Lindisfarne Press, 1992), p. 129f.
29 See last chapter in Apuleius, *The Golden Ass,* Richard Aldington trans. (1566).
30 'The divinities are the beckoning messengers of the godhead. Out of the holy sway of the godhead, the god appears in his presence or withdraws into his concealment. When we speak of the divinities, we are already thinking of the other three along with them, but we give no thought to the simple oneness of the four.' Martin Heidegger, 'Building, Dwelling, Thinking,' in *Poetry, Language, Thought* (New York: Harper & Row, 1971), p. 149.
31 Heidegger, *Op cit.*
32 Shakespeare, *The Tempest*: II.i, 177–9.
33 The actual name given to the project at NASA, using plutonium-fuelled rockets to explore deep space.
34 de Nerval, Gerard, 'Golden Lines,' in Robert Bly, trans., *News of the Universe* (San Francisco: Sierra Club Books, 1980), p. 38.
35 Quoted in Svoboda, Robert, *The Greatness of Saturn* (Sadhana Publications, 1997), p. 16.
36 Paris, Ginette, *Pagan Meditations* (Dallas: Spring, 1986), pp. 175–7.
37 *Ibid.*
38 Sappho, Poem 16, in *Poems of Sappho* (translated from the Greek publication).
39 Yeats, W. B., 'The Cat and the Moon,' in *Collected Poems* (New York: Macmillan, 1951), p. 164f.
40 Blake, William, 'The Tyger,' from *Songs of Experience.*
41 Shakespeare, *The Merchant of Venice,* IV.i, 184–86.
42 Munch, Edvard, as quoted in Noel Cobb: 'The Morbid and the Beautiful,' *Sphinx* (The London Convivium for Archetypal Studies, 1988), p. 46.
43 Blake, William, 'The Marriage of Heave and Hell.'
44 Lawrence, D. H., 'The Nightingale,' in *Sketches of Etruscan Places and Other Italian Essays* (London: Penguin, 1999), p. 212.
45 *Ibid,* 'The Painted Tombs of Tarquinia,' p. 6l.
46 Shakespeare, *King Lear*, V.iii.257ff.
47 Hillman, James, *Emotion* (Evanston: Northwestern University Press, 1992), p. 265.
48 Keats, John, 'Lamia,' line 230.
49 Sardello, p. 24.
50 Dante, 'Paradiso' *La Divina Commedia,* Canto 33, Line 145.

7. The Russian Spirit of Place

CHERRY GILCHRIST

How do Russians view the spirits of land and sky? Some of Russia's most ancient heritage can still be found in Siberia today, where animism and shamanism is the predominent religion, revealing the life of the landscape and sky as a coherent cosmology. Different spheres of life interact, explaining the link between human life, the sky and the spirit world. Personal experience of shamanic ritual in Siberia sheds light on this, and on possible ways of viewing our own landscape to re-ensoul it. In European Russia too, there is a rich heritage of belief in the living forces of nature, the elements and the heavens. Drawing on study, research and personal journeys, this paper shows how the spirits of forest, river, frost, snow, sun and wind are colourfully depicted, how they may be revered or feared, tricked or negotiated with, but always respected. This also plays a part in shaping artistic traditions, with the belief that new artistic inspiration can literally float down from the sky on a feather of the Firebird. Myth, art, soul and ideas of beauty are all tied into the Russian view of their land and sky. The talk was illustrated with a remarkable and unique collection of slides from Russia and Siberia, a few of which are included here.

It is my own experiences in Russia which have shaped this perspective on the Russian Spirit of Place. I could not have assembled these reflections from books alone; it was the real life contact with the country itself, and seeing at first hand what the spirit of place means in terms of creativity and culture, that have given me what I consider to be real and genuine insights.

The idea that there is such a thing as spirit of place is very much a central element of Russian culture; it seems that this has always been the case, as far as history can tell us, and it is certainly still true today. We can find this concept embodied in the Russian traditional art forms and way of life, as I shall show. As will also be made clear, a strong element of it is based on a sense of the relationship between Earth and sky. It can be argued that its underlying roots are in shamanism, the animistic and indigenous cosmology which has now evolved into different forms in the main areas of Russia, but which can still be found in its old forms in Siberia.

The Shamans of Siberia

In the summer of 2004, I set out on what was to be my fifty-sixth visit to Russia. I had been travelling to and from Russia since 1992, to study traditional art and craft there and to bring it back to the UK for exhibitions and displays. However, I had never visited Siberia before; it had been a long-cherished wish, and finally all the ingredients came together and I was able to spend some time in Tuva and Khakassia, two provinces close to the Mongolian border to the south, and adjoining the Altai region to the west. It was August, and while England suffered from rain and storms, we basked in temperatures of around twenty-five degrees centigrade. The landscape was a striking mixture of open steppe lands, broken up by round, rolling hills and jagged mountain ranges capped with coniferous forests. In between were green hillside meadows, some carpeted with alpine flowers and graceful larch trees. Most people think of Siberia either as a snowy waste or as a monotonous stretch of plain and tundra; but although these southern areas could plummet to minus forty degrees centigrade in winter, their beauty during the other half of the year was breathtaking.

In much of Siberia, shamanism is the predominant religion. Although it was largely suppressed in Soviet times, it is now making a very strong comeback, and in Tuva and Khakassia, it is the primary belief system, followed by Buddhism in second place. Shamanism is principally an animistic religion in which spirit and spirits are known to inhabit the world around us. It is the opposite of the dead, mechanistic universe proposed by Newtonian physics; it is not a blissful, idyllic vision of life, however, as spirits of animals, mountains and departed ancestors can also be angry and vengeful, as well as wise and helpful. The shaman acts as the intermediary between humans and spirits, gaining foreknowledge, acting as a channel for healing and as an agent of empowerment. It is beyond the scope of this paper to give a thorough analysis of shamanism, or to discuss the different arguments as to how it should be defined; I prefer to present here some of the elements which relate to the spirit of place, and some of which, as we shall see, have been carried through into traditional culture elsewhere in Russia. [1]

One of the first features of the landscapes that strikes you when you visit southern Siberia is that there are shrines everywhere. Gaunt branches, thrust into small cairns of stones, garnished with coloured ribbons and rags tied to them, look themselves like strange skeleton spirits. Every significant place has its own shrine: hilltops, rivers, rocks, wells, and, in our case, the visitors' yurt camp where we stayed. Siberian cosmology does, however, differentiate between places which are especially sacred, and those which have little significance. Although animism means that the world around is perceived in general as alive, not every single feature or object is important; in a mound of pebbles, for instance, only certain stones will be seen as embodying a powerful spirit. The landscape, therefore, has its spiritual contours too, and the peaks and 'hot spots' of these can be marked with shrines.

The ribbons and rags represent prayers, offerings and wishes. I never had such plentiful opportunities to make wishes as when I was in Siberia. At one shrine we visited, situated close to a cult stone, everyone in our small group was presented with a red ribbon and invited to tie it to a twig on the shrine's branches while making a wish. As I began to tie my ribbon, I realized that the making of a wish was not necessarily a simple affair. The shrine was acting as a witness to my act, to my own integrity, and my clear-sightedness, or perhaps my foolishness, in defining what I wanted. Did I really want it after all? The sharp shock of seeing the possible consequences of fixing my desire on a certain goal brought its own insights, which reverberated and remained with me for weeks to come. The shrine can therefore be a test of one's own honesty and commitment.

Siberian cosmology is at base a threefold system, with our immediate, earthly world in the centre, a world of spirits and sky above, and an underworld below; sometimes referred to as settlement, sanctuary and cemetery respectively.[2] This fundamental division into three is common to shamanic or animistic belief systems in other parts of the world. I will also give examples later of how it has remained at the heart of Russian traditional culture. But the Siberian shamanic view builds on this basic concept to create a world view of extraordinary complexity. There are often not just three divisions of the universe but seven, nine or sixteen worlds, according to the type of shamanism, and the perceived number of levels may be changed according to the ritual that is being practised.[3] It is a fluid cosmology, highly structured, but with a structure that can be viewed in different ways according to place and purpose. To add to this complexity, even spatial dimensions are not fixed: vertical and horizontal may be interchangeable, and a river perceived as a sacred river of life may be seen, for instance, both as flowing from east to west, and from above to below. Reading up on Siberian cosmology is a disorientating experience, with dizzy perspectives that are perhaps the equivalent of our own attempts to struggle with the modern notion of the relativity of time and space.

This cosmology is also an intrinsic part of the landscape, and thus itself contributes to the sense of a spirit of place. The entrance to the underworld, for instance, may be that specific cave which you can glimpse on a hillside on the other side of the valley. Here the spirits of the deceased, so the teaching goes, may congregate, having been led there by an elk or significant animal spirit. Literal features of the landscape can thus be entry points to another world, and there is sometimes a mirroring between one world and another; in Tuvan shamanism, for instance, anything which is broken in the everyday middle level world is considered to become present in its unbroken state in the underworld — plainly a great consolation if something precious is broken, which will then be available for the departed ancestors to enjoy.

But at the centre of the shifting perspectives of Siberian cosmology is the axis of life, linking above and below. As mentioned above, even notions of vertical and horizontal may switch, but by and large, the axis running between the two poles of creation is a steady concept.[4] Sometimes this is represented by a sacred mountain in the landscape. It may also relate to a cult stone, such as the *Starushka* or 'Old Lady' stone at the shrine where we tied our red ribbons. In Tuva and Khakassia, many of these stones, dating from the Bronze Age or earlier, are still in common use. Women trek to the *Starushka* and make offerings to her, seeking her help particularly in cases of infertility. At another site known as the 'White Stone,' we tried out the local practice of walking three times sun-wise around the stone and then holding it gently for a short time — being close to it for too long is said to be dangerous because the energy is powerful. As we were in Russia, I had no hesitation in asking our guide, the leading archaeologist of the region, if any scientific research had been done on the stone's special qualities. In Western Europe this would often be considered a crackpot question, but in Russia, where there is an avid interest in psychic studies and the paranormal, this was perfectly reasonable. Certainly, he told us, teams of scientists had done just that and concluded that the stone had an unusually potent energy field around it.[5]

By now it will be clear that the world axis is not just found in a single permanent location in the shamanic cosmology, but can be present in different places or called into being for the occasion. One common way of representing the axis is by creating a 'World Tree' (for instance using a small birch tree) or ladder, which a shaman in trance state ascends to visit higher realms.[6] He or she may be helped in this 'flight' by a spirit guide that takes the form of an animal or bird, such as a horse, eagle or crow.

I took the opportunity to have a private session with Herel, a Tuvan shaman in his thirties. Nowadays, many of the shamans work in 'clinics,' in a designated room hung with animal heads and skins, bones, ribbons, whips, drums, patterned cloths and other items of power. Herel offered healing and divination, whereas his shamaness wife specialized in treating women's ailments. They had five children, and expected their youngest girl to become a shaman too; the signs were already there: 'As she was being born, the weather was changing constantly, from sun, to snow, to thunder.'

I sat on a bearskin rug, feeling some trepidation. Herel gave me an immediate diagnosis of the problem: he said that he saw something heavy in my left shoulder, connected with certain troubles that had oppressed me. He would, he said, take out this heaviness. In two months' time, he predicted, I would be looking younger and feeling much happier! He lit some juniper brushwood, and held each of my hands in turn over the smoke, before passing the burning juniper itself around my body. I was asked to close my eyes and fold my hands in prayer position. 'Relax,' he said. 'Don't be afraid.'

Herel had unbraided his hair, a sign that the forces of spirits and magic could be unleashed, and he was now wearing a ceremonial eagle's-feather headdress, and a shaman's loose black leather coat whose fringes flicked across me every now and then in a feathery touch.[7] He circled around me, drumming and chanting. To my surprise, it all felt completely natural; with my eyes firmly shut, I felt bathed in the sound as though it was coming from different places at once. The sound intensified as he began to use cymbals, and later he moved into a different phase of chanting.

Visions filled my head, most notably dancers with demon or lion masks, and an eagle which sat calmly by my side, communicating with me through a wordless intelligence. Shock hit my body as I was thumped hard on the shoulder with something that I later found out was a bear's paw, but the use of the whip, a standard shaman's instrument, was stimulating and pleasant. Perhaps more of an outrage to my Western sensitivities was the moment when he lifted my T-shirt away from my neck and spat copiously down my back!

I finished the session dazed, but at ease. Herel gave me his prognostications for the year ahead and gave me a kind of spirit pouch for good fortune, in the form of a little cloth bag stuffed with grain, to be hung up high in my bedroom by its braided thong, and requiring regular 'feeding' with oil or melted butter.

Then he advised me to make contact with the spirits of the hills and rivers where I live. 'If you have them,' he added. At first, I found this strange, although I was perfectly at easy with visiting shrines and sacred mountains and shamanic ceremonies in Siberia. But in England? On the tame hills above Bath, could I possibly find the equivalent spirits of place? I decided to be open to these possibilities. But where was our own spirit of place? Perhaps it had been overlaid by our so-called civilized outlook, and I suspected that this was not so much the rational, scientific viewpoint, but more the romanticized, eighteenth-century view of the countryside, which encourages us to see 'nature' as a sympathetic medium in which to experience our own personal feelings, while marvelling at 'her' beauty. The Siberian spirit of place, it seemed to me, was something more raw, vital and powerful.

For several months after my return, I walked the hills above the city of Bath each morning, and tried to pick out what might be powerful points or constellations of energy in the landscape. I stood under trees, and attempted to sense the different nature of each one, and to find out how this might connect to me as a human being. Gradually I realized that this was not a kind of sentimental endeavour; connecting with the spirit of place was about extending one's awareness, and about being receptive to other forms of intelligence and life. Even though we will always clothe such interpretations in our own cultural imagery, I discovered in Bath that it is still possible to find spirit of place nearer to home rather than just in exotic, unfrequented Siberian landscapes. Even though, I suspected, some of our spirits of place might be by now just a little weary and trampled over!

The house as microcosm

Much of the rich realm of Russian folk culture probably derives from the original animistic or shamanic belief system, which is thought to have extended over the whole of northern and central Russia in ancient times.[8] The tripartite cosmology of sky, Earth and underworld, so prominent in Siberia, is also identified in many fairy tales and folk art motifs, and is embodied in the plan on which Russian wooden houses are usually built.[9] Each house was traditionally seen as a microcosm of the cosmos, and although awareness of this symbolism may have waned, practically every village house is still built in a similar way today. The ground floor, which is often an open-plan heated living and sleeping area, represents the earth and the everyday world that we know. The cellar, usually reached through a trapdoor in the floor, and used primarily to store food for winter, is the underworld where the spirits of the dead ancestors may reside. And the attic, unheated and therefore used largely in summer, is the place of the sky spirits. Horses, sun symbols and peacocks may be carved on the gable ends to symbolize the protection of the celestial forces, and young girls climb eagerly up the ladder into the loft space to practise divination and play magical games under their auspices.

Other features of the household play their cosmological role. The Red Corner, traditionally situated in the main living room opposite the doorway is the holy place, where the family icon is placed on a high shelf, draped with a white linen 'towel' (a long band of cloth) embroidered in red. Red has the significance of beauty in Russia, with the words for red *(krasni)* and beautiful (*krasivi*) stemming from the same root. The icon, ideally painted under strictly prayerful conditions in terms of Orthodox belief, is considered as a medium of divine grace, which receives the prayers of the family and may watch over them and bless them in return. During the anxious wait for news of survivors from the *Kursk* submarine disaster in August 2000, an old lady sobbed as she told a television reporter that their icon had just fallen from its shelf, and that this was a terrible omen for her grandson trapped on board.

The Orthodox religion, which was adopted in Russia in the tenth century, has by and large co-existed peacefully with the indigenous belief system, which by then was probably an evolution of the earlier shamanic practices, and which is often described as a nature religion. Almost until the present day, Russia was known as the country of two faiths, and perhaps it is not therefore surprising that we find this polarity of Christianity and paganism embodied in the cosmology of the home. While the Red Corner represents the Christian pole, the bathhouse, usually outside in the garden, is often considered as its opposite pole, the repository of earlier beliefs and customs.[10] Here a bride spent her wedding eve with her girlfriends, with a local sorcerer conducting the preparatory rites, and here too babies were born and then sometimes presented to the stars outside by the midwife.

In the home itself stands the sizeable stove, which can perhaps be considered as the reconciling force between the two religions. It is known as 'mother,' and is the provider of warmth, cooking and drying facilities, and sometimes a bed for the night too on its flat upper surface. Whatever the religion, no one can do without that basic comfort and sustenance from the welcoming, maternal stove. External protection is given by the carved wooden fretwork around the windows, mentioned earlier, which is said to repel evil forces trying to enter the home. (Another version of its purpose is said to be to frame young girls attractively as they sit and spin, thus increasing their chances of getting a husband).

Spirits of nature

In terms of the old beliefs, the Russian landscape is peopled with nature spirits, *Leshi*, the master of the forest, is a shape-changer who can appear as a troll-like figure, or who can loom 'as tall as a belltower,' according to one eyewitness report, when he acts as a guide to the animals and birds in his care, in this case leading them to safety from a forest fire. One must never whistle in his presence, or stay the night in the woods without his permission; offenders will be led astray by *leshi* and become irrevocably lost, something that can happen all too easily in the vast stretches of Russian forest.

Vodyanoy is king of the river, a merry carouser, whose presence can be felt by waves on the water; choppy ones mean that his guests are having

Figure 1: Vodyanoy, *King of the river, painted on a lacquer miniature. By the artist Shatokhina from Palekh (from the author's own collection).*

a good party down below, and a single standing wave that vodyanoy is galloping along the riverbed in his carriage. He may be on the hunt for black cows, for which he has a passion. Rusalki, mermaids of the streams, leave their watery homes and prowl about the forest looking for appealing young men to lure into the depths, and in winter Morzoko (Father Frost) and Snegurochka (Snowmaiden) represent the fierce elemental force of cold and the delicate beauty of snow respectively. The form that nature spirits are known in now may well have evolved from earlier, fully-fledged deities, which included the sun god, Yarilo, and Perun, the lord of thunder. Nature spirits are more localized — each wood has its leshi, for instance — and more open to interaction with humans, appearing both as characters in fairy tales and in country folk's accounts of encounters with them. Some of the older deities became Christianized so that, for instance, Perun became associated with Elijah, who also had a habit of riding through the sky in a chariot. Elijah is thus considered as a fearsome character in popular Orthodox belief, one who casts thunderbolts and smites evil doers, as Perun used to do.

Spirits are also present in the microcosm of the home. Every home in Russia is said to have its *domavoy*, or guardian spirit. However, like many of his kind, he is not entirely benevolent and is given to waking up at midnight and banging about the house. He lives behind the stove, perhaps representing another internal polarity in the cosmology of the home, as a male prankster contrasting with the stove's maternal warmth, and he has to be kept in a good humour for the well-being of the family, often with gifts of porridge. Again, like many nature spirits, he is a shape-changer, who may often be seen as an old man wearing a shaggy hat and a red sash, but

Figure 2: Lacquer miniature of Snegurochka, *'Snowmaiden,' from Kholui (from the author's own collection).*

who can just as easily appear as a horse, snake, hen, magpie, goat, cow or a fir tree within the territory of the homestead.

Beliefs in nature spirits are now considered by some to be merely interesting folkloric data of a largely bygone era, but to say that they have vanished would be far from true. They are painted with reverence by the Russian lacquer miniature artists, and recounted in all kinds of tales and legends. And they are still perceived at first hand, not only by country folk but by more sophisticated inhabitants of the Russian landscape as well. On the two occasions that I have visited Kiji island, in Karelia, I have asked the guides there about their personal experiences. This is an island in the far north, and certainly remote in its way, but much visited in summer by tourists for its famously beautiful wooden churches, and the guides are usually students or experts in history or folklore who work out there for a season. One young female guide reported to me that she senses the presence of a spirit, which she identifies as a *domavoy*, each time she has to stay alone in one of the old village houses. Sometimes she hears a knocking sound, or a melody being played. Another told a tale against her young male colleague, who declared himself a sceptic in such matters. However, he failed to leave a little bit of soap and a fir twig as an offering for the *bannik*, a rather malevolent spirit who inhabits the bathhouse. On his next visit, he banged his head coming into the bathhouse, dropped his glasses and trod on them, breaking them to smithereens. Now he is a convert, who always leaves a suitable gift for the *bannik*.

The Firebird: the quest, art and the spirit of place

The Russian Firebird is the symbol of inspiration; as a sky spirit, she gives to the Earth her revelation of blazing light, often initiating quests in fairy tales and in real life seen as the source of new artistic endeavour.[11] She appears in many Russian tales, and it is worth mentioning at this point that fairy tales in Russia are taken seriously; they are known and loved by people of all ages, and are considered as a profound element of the cultural heritage. As one artist put it, 'They carry the wise thoughts of poor people.'[12] One of the finest examples of the Firebird quest is found in the well-known tale of *Prince Ivan and the Firebird*, in which the young prince sets out in search of the Firebird after he discovers her at night his father's orchard, stealing the golden apples from the trees. He manages to grasp one feather from her tail before she escapes, a feather

Figure 3: Lacquer miniature of the Firebird. By artist Pyotr Mityashin of Kholui (from the author's own collection).

whose light is so brilliant that he cannot rest until he sets out in pursuit of her. Ivan undergoes many ordeals; he even suffers death at the hands of his jealous brothers, but his trickster friend Grey Wolf despatches two ravens to the otherworld to fetch the Waters of Life and Death to bring him back to life again. After many adventures, Ivan captures the Firebird and returns to the palace, having also gained a new horse with a golden mane and tail, and a beautiful princess for his bride. Although he has consistently disobeyed instructions and ignored good advice, he comes through it all, returning in the end he to the place whence he came, a wiser and wealthier man.

The Firebird can therefore inspire a quest involving a challenging journey which eventually takes the hero full circle back to the place of origin. This place, and the hero's relationship to it, will never be the same again after the trials and revelations of that journey, and as well as any material rewards gained, love, wisdom and enlightenment are often the real prizes.

The other intriguing belief about the Firebird in Slavic Russia is that she is also an inspirer of art.[13] 'Wherever a feather of the Firebird falls to earth, a new artistic tradition will spring up,' goes the saying. They also say that such a feather fell upon a country area known as Khokhloma, not far from Nizhni Novgorod, inspiring the creation of lacquered and painted wooden ware. This is a craft form that is well-known throughout Russia, and most families have a piece or two in their kitchen or living area — a spoon, bowl or flower vase, for instance. The Firebird, it is believed, was directly responsible for giving local craftsmen the idea of decorating wooden platters and cups with stylish designs in black, red and gold, and then lacquering them over to produce a durable finish.[14] She may also have dropped a feather or two on other villages and regions where local craft industries sprang up; there are many regional art forms in Russia, where villagers have often worked away on crafts during the long winter nights. Russian people have a natural tendency towards art and decoration, and even their wooden houses are adorned with beautifully carved lacy fret-work windows. Other prominent regional crafts are metal trays painted with flowers from Zhostovo, carved toy bears from Bogorodsky, woven shawls from Pavlovsk Posad, and nesting dolls from Sergiev Posad.

With Khokhloma ware, the golden and red, feathery, delicate swirls of the painted patterns may remind us of that initial Firebird's feather that drifted down to earth, though a closer look may reveal that the motifs are often drawn from the natural world of berries, flowers and ferns. The colours create a vibrant and even fiery effect, so that perhaps it would not be too far-fetched to see here an art that springs from a combination of celestial inspiration and earthly beauty. As with the other craft forms, the mythic dimension and symbolic attributions are taken very seriously; the Khokhloma colours are interpreted as black for compassion and suffering, red for energy and beauty, and gold for hope and eternal life.[15] In the decorated dolls of Semyonov, not too far away, hens and birds stand for happiness, flowers for beauty, and the women's headscarves for protection. It pays to give serious attention to Russian crafts; although they

Figure 4: Khokhloma ware

Figure 5: Painting dolls at Semyonov

may have humble, mundane uses, they are created with consummate skill and artistry, and are imbued with rich symbolism, some of which goes back to ancient times.

When an artistic or craft tradition becomes established in a particular place, then a strong sense of the spirit of that place builds up. The artists' love of their village merges with their pride in the art itself. I discovered this on my own quest within Russia, which itself began with the Firebird. It was in 1992, while visiting St Petersburg for the first time, that I first set eyes on Russian lacquer miniatures, arguably the finest traditional art form produced in the country. These are generally painted on small papier-mache boxes, capturing scenes from fairy tales and legends, landscapes and calendar customs. The exquisite, highly detailed work, intensified by its miniature form, and painted in glowing colours on a deep black background, captured my imagination powerfully. I felt that there was something of the soul of Russia embodied in this. One miniature in particular enchanted me: a scene from the story of *Prince Ivan and the Firebird.* It showed the prince reaching up to seize the tail of the Firebird, just as she soared upwards in a blaze of light, leaving the golden apple tree behind. I admired this painting immensely, but I did not yet know anything about the image it represented or who had originally created it.[16]

Over the next few years, I located the four different villages where this art is practised: Kholui, Palekh, Fedoskino and Mstiora. Each has its

Figure 6: The Church at Kholui

own style of art, and each of the four villages has its own character and atmosphere, as well as a workshop, training school and museum with breathtaking displays of miniatures.[17] My own first visits to the villages had a sense of the mythic about them. The experience was heightened by the fact that it was not easy, even in a supposedly free Russia after the fall of Communism, to visit country villages. (Three of the lacquer miniature villages are about 250 miles east of Moscow, and the fourth just north of the capital.) To make the trip needed organization, persistence, and most important of all, people who could help you and introduce you to artists and studio directors when you were there.

By the time I reached Kholui and Palekh for the first time, I had fallen in love with the art, and regarded my trip as something of a pilgrimage. I could not imagine who these semi-divine beings were who created such magical miniatures. Of course, as it turned out, they were simply people — warm-hearted, sensitive, intelligent artists whose magic perhaps was that they combined the practice of fine art with an earthy, traditional life, where planting the season's potatoes might be just as important as putting the finishing touches to delicate gold ornament on an exquisite miniature. The villages themselves were particularly meaningful to me personally, as I found a living culture there that bore traces of our own long-vanished way of life.

There were meadowlands without fences, full of wild flowers, as well as cows, goats and chickens wandering freely around the grassy village streets. The wooden houses, all painted different colours, charmed me, the roofed wells were something out of a fairy tale, and the broad river beckoned you to swim in its slow-moving waters, studded with yellow lilies. I shocked some of the friendly female artists in Kholui at lunch one day by telling them in all innocence that I loved their village because it reminded me of medieval England. They did not take it as a compliment. But all artists in all four lacquer miniature villages are immensely proud of their art, and the place which gave birth to it. Each village vigorously affirms its own identity, the superiority of its own spirit of place, even while paying respectful tribute to their colleagues in the other three villages. The character of each village becomes enshrined in the art too, since the graceful white church of Kholui appears in many of the miniatures' landscapes, as do the winding river of Fedoskino, the fair at Mstiora, and the scenes of mushroom and berry picking at Palekh. The Firebird is present too, again as a symbol of inspiration; as well as the Firebird's popularity as a theme in lacquer miniatures, she is stamped as a trademark on boxes from Palekh and Kholui.

My visits to the villages were often idyllic, though as time went on I learnt that the compelling spirit of place is not just about picnics in the forest and parties in the snow. I learnt that Russians' merry-making, as well as their permanent quest for art and beauty, is often a counterbalance to the harsh life in a social climate where medical resources are poor, where early death is a distinct possibility, and the vagaries of changing times create terrible financial pressures and lack of security. But I also recognized that the mixture of joy and tragedy with which their lives was so often marked, was also distilled into their work, and it coloured too the spirit of the place in which they lived.

I had thus followed the Firebird to its source; finally, in an artist's house in Kholui, I found the original version of that one Firebird box whose picture I had first seen copied on a miniature in St Petersburg. It was an extraordinary moment, coming face to face with the image which had inspired my quest several years earlier. When the artist showed it to me, I knew that my journey had reached its furthest point, and that the second half, the return towards home, was about to begin. Later, in 2004, after twelve years of visiting Russia, I realized that the journey in terms of travelling was now over, but that now a new phase had begun, which would involve distilling my own understanding of those visits and taking stock of what I had learned. Russia's spirit of place is not my own spirit, or my own place, and there were times when that distinction was sharp and painful. But I was fascinated and enchanted by it, and ultimately I have been greatly enriched by a close connection with it.

The spirit of Russia

Russia is an enigmatic and mysterious country, which has survived many harsh regimes and political upheavals. In my view, it cannot be understood simply by reading the history books or watching reports on the media. One can only touch on the enduring spirit of Russia by studying the relationship of its people to land and sky, and by becoming absorbed in the culture that this generates, whether it is shamanism or the fine art of a lacquer miniature. People's perceptions of this relationship can be responsible for fashioning the construction of the home, for evoking a landscape inhabited by spirits, and producing a creative and colourful range of craftwork. It is this spirit, imbibed at least to some degree by practically every visitor to Russia, that leaves most foreign visitors feeling uplifted, enthused and energized when they return home, despite the sometimes grim political and urban conditions they may encounter. I have witnessed many such tourists returning from their trips, and talked to many people who have lived or worked in Russia over the last few decades, and this is almost without exception the case. The Russian spirit of place is perhaps the prime element that ensures the continuation of Russian culture, and the survival of Russian people during difficult times.

Notes

1 See the Bibliography for details of some useful publications on shamanism in general, and Siberian shamanism in particular.
2 For further details of these terms, the multiple levels of the different worlds in Siberian shamanism, and the perception of spatial directions, see in particular the paper by Kosarev, M. F., 'The System of the Universe in Pagan Siberian Indigenous Peoples,' *Astronomical & Astrophysical Transactions,* 17:6 (1999).
3 There are various regional categories of shamanism in Siberia, each with its own differing practices and cosmology. Examples of these are the well-known Tungus (or Evenk) shamans, the Buriats, Samoyeds and Yakut. For a general survey of these and other types of shamans, see Stutley, M., *Shamanism: An Introduction* (London: Routledge, 2003).
4 See for instance Diószegi, V., and M. Hoppál, *Shamanism in Siberia* (Budapest: Akadémiai Kiadó,1978), p. 32.
5 The specific information given was that the 'White Stone' has a high charge of radio-active energy coming from underground, which has been measured three

times by scientists. The first expedition came in 1997 from St Petersburg, the second from Tomsk, and the third from Abakan in Khakassia.

6 Diószegi, *Op cit,* pp. 117f.

7 Unbraiding of hair and the untying of a belt are also considered to be powerful actions in central Russia, symbolizing the opening of the way to the potentially chaotic influence of magic. This might happen, for instance, in the bathhouse or the attic, the places in the homestead which are most strongly associated with spirits. See endnotes 9 and 12.

8 See Ryan, W. F., *The Bathhouse at Midnight: Magic in Russia* (Stroud: Sutton Publishing Ltd., 1999), p.11 [hereafter *Ryan*].

9 See Krasunov, V. K. (ed.), *Russian Traditions* (Nizhni Novgorod: Kitizdat, 1996), p. 121, and Haney, Jack V., *An Introduction to the Russian Folk Tale* (New York & England: M. E. Sharpe Inc, 1999), pp. 48f, for the tripartite cosmology of the house, and Hilton, Alison, *Russian Folk Art* (Bloomington and Indianapolis: Indiana University Press, 1995), p. 177 for examples of three levels in folk art.

10 *Ryan*, p. 58, quotes a contemporary book on magic which has a chapter on becoming a *koldun* (a traditional type of sorcerer, such as used to be found in every Russian village), posing the question as to the best place to start training. The answer is: 'Anywhere you lie. But best is a bathhouse … at midnight.'

11 The Firebird, always female, may be related to the Simurgh, the divine bird of light found in Central Asian mythology.

12 Nikolai Baburin of Kholui, interviewed by Cherry Gilchrist in 1997.

13 I use this phrase to describe what we might call 'ethnically Russian' Russia, chiefly the western and central areas, and to represent Russian culture approximately from the tenth century onwards, the period when Slavic domination was established.

14 The 'gold' is actually created by applying powdered metal of a silver colour, which then turns golden when heated. This technique is one of the chief features (and at one time, secrets) of Khokhloma craft.

15 Interview with director of Khokhloma workshop and museum, 1999.

16 Lacquer miniature compositions are created first by one artist, but may subsequently be copied or adapted by other artists. A 'repertoire' of images thus builds up, much as it does in icon-painting; the emphasis is not on complete originality every time, but on staying true to the spirit of the art itself.

17 For the history and technique of lacquer miniature painting, see Gilchrist, Cherry, *Russian Lacquer Miniatures* (Bristol: Firebird Publications, 1999).

Bibliography

Shamanism

Diószegi, V. and M. Hoppál, *Shamanism in Siberia* (Budapest: Akadémiai Kiadó, 1978).

Eliade, Mircea, *Shamanism: Archaic Techniques of Ecstasy* (London: Penguin Arkana, 1989 [1964]).

Gorbatcheva, V.. and M. Federova, *The Peoples of the Great North: Art and Civilization of Siberia* (New York: Parkstone Press, 2000).

Harvey, Graham, *Shamanism, A Reader* (London & New York: Routledge, 2003).

Hutton, Ronald, *Shamans: Siberian Spirituality and the Western Imagination* (London and New York: Hambledon & London, 2001).

Kenin-Lopsan, Boraxoo M., and E. Taylor, 'Tuvinian Shamans and the Cult of Birds,' *ReVision* 19.3 (1997), p. 33.

Kosarev, M. F., 'The System of the Universe in Pagan Siberian Indigenous Peoples' *Astronomical & Astrophysical Transactions* 17: 6 (1999).

Riordan, James, *The Sun Maiden and the Crescent Moon — Siberian Folk Tales* (New York: Interlink Books, 1989).

'Sarangarel' (Julie Ann Stewart), *Riding Windhorses* (Vermont: Destiny Books, 2000).

Stutley, M., *Shamanism: An Introduction* (London Routledge, 2003).

Vitebsky, P., *The Shaman: Voyages of the Soul* (London: Duncan Baird, 1995).

Russian traditional art and culture

Billington, James H., *The Icon and the Axe* (New York: Vintage Books, Random House, 1970).

Gilchrist, Cherry, *Russian Lacquer Miniatures* (Bristol: Firebird Publications, 1999).

Haney, Jack V., *An Introduction to the Russian Folk Tale* (Armonk, New York & England: M. E. Sharpe Inc, 1999).

Hilton, Alison, *Russian Folk Art* (Bloomington and Indianapolis: Indiana University Press, 1995).

Ivanits, Linda J., *Russian Folk Belief* (Armonk, New York & England: M. E. Sharpe Inc., 1992).

Krasunov, V. K. (ed.), *Russian Traditions* (Nizhni Novgorod: Kitizdat, 1996).

Milner-Gulland, R., *The Russians* (USA: Blackwell, 1997).

Phillips, C., and M. Kerrigan, *Forests of the Vampire* (London: Duncan Baird Publishers, 1999).

Rozhnova, P., *A Russian Folk Calendar* (Novosti, Moscow, 1992)

Ryan, W. F., *The Bathhouse at Midnight: Magic in Russia* (Stroud: Sutton Publishing Ltd., 1999).

Warner, E., *Heroes, Monsters and Other Worlds from Russian Mythology* (Eurobook, 1985).

Zenkovsky, S. A. (ed.), *Medieval Russia's Epics, Chronicles and Tales* (New York: Meridian, 1963).

Russian Folk and Fairy Tales

Afanasiev, Aleksandr, trans. Norbert Guterman, *Russian Fairy Tales* (New York: Random House, 1973).

Gilchrist, Cherry, *Prince Ivan and the Firebird* (Bristol: Barefoot Books, 1994).

Maxym, Lucy, *Russian Lacquer Legends* Vol. 2 (Hisksville, NY: Siamese Imports Co. Inc., 1986).

8. A Dimensional Model for the Relationship of Consciousness and Cosmos: Mathematical Abstraction versus Conscious Experience

ROBERT HAND

In the Sefer Yetzirah, *an early Kaballistic work, there is a description of the relationship among a group of ten polarities: North-South, East-West, Up-Down, Good-Evil, Beginning-End. This paper will examine the apparently dimensional nature of this framework and how it reveals a difference between our mathematical/abstractive approach to such matters and an ancient Pythagorean/qualitative approach. In our mathematical/abstractive approach we may have lost the ability to visualize and model the relationship among various levels of consciousness.*

In this chapter we are going to explore a possible way of modelling consciousness, as well as different states of consciousness, in terms of dimensions and dimensionality. However, to do this we must look at dimensions and dimensionality somewhat differently than is done in modern mathematics. In modern mathematics we have an extremely abstract system based completely on logical relationships and pure extended quantities or magnitudes, that is, quantities that have no qualities whatsoever except that of being magnitudes extended in space (whatever 'space' may mean).[1] In the ancient world, mathematical entities also had qualities of other kinds in addition to logical relationships and extended magnitude. Logical relationships, extended magnitude and these other kinds of qualities were all examined and considered not only meaningful but also the proper study of mathematics. This paper will argue that by bringing back into consideration these other kinds of qualities — qualities that could be experienced directly by the senses, or visualized directly by the mind — we can begin to model consciousness and even to understand something of what the ancients have told us about divinity.

Pythagorean numbers versus modern numbers

To introduce the basic idea that we are going to be working with here, we need to look briefly at the ways in which the Pythagoreans, in particular, looked at whole numbers, and contrast that with the manner in which modern mathematicians look at them. We do this because these ways of looking at numbers by Pythagoreans are more well-known than some of the material we will be looking at below.

While the Pythagoreans did work with extended magnitudes in space in their geometry, it is in their use of whole numbers, or integers,[2] that we can see the manner in which they differed from modern mathematicians.[3] The modern way of dealing with a series of integers — 1, 2, 3, 4, 5, 6, 7 etc. — is to see only the purely numeric aspect of each number. The number two differs from one in having twice as many elements to be counted. Six also has twice as many elements as three. If we add one and three we get four, and again the only difference between four and three is the number of things being counted. The point here is that there is no other difference among these several numbers. There is nothing about three-ness that makes it different from four-ness aside from the pure abstract number. Whole numbers or integers are simply aggregations of units, which differ only in their respective numbers of elements. There is nothing else that makes them different.

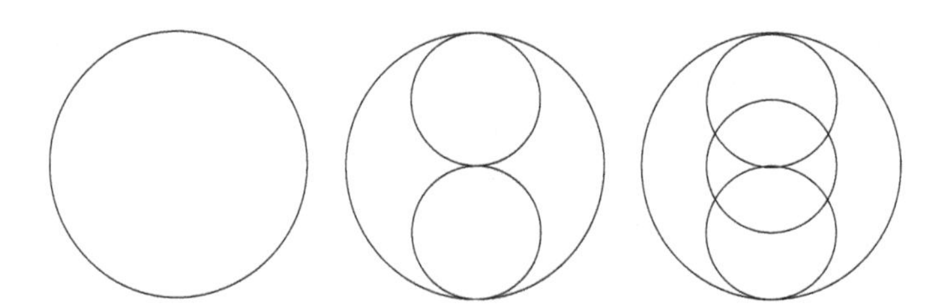

Figure 1: Unity, Duality and Harmony

The Pythagoreans saw whole numbers differently. Among other possible distinctions they saw two categories of distinction in particular. In figure 1 we see the first way in which qualities can differentiate numbers from each other.[4] The number one is viewed not only as singularity, but as wholeness, completeness, the All. This is to be seen in the tendency of many early CE classical philosophies to call the highest aspect of divinity 'the One.' The classic example of this is of course in the work of Plotinus

(205–270 CE) and the other Neoplatonists. The One is both singularity and wholeness, a transcendent wholeness which includes and yet transcends everything.

Two, on the other hand, is more than a pair of singularities, or units as they later came to be called. It is the principle of opposition, conflict and polarity. This aspect, or quality, of two has been preserved in astrology where the division of the circle into two parts, which occurs when two planets are directly opposite each other in the zodiac, is called an opposition. The opposition is described in astrology as having all three of the qualities just described as attributes of the number two.

Limited	**Unlimited**
Odd	Even
One	Plurality
Right	Left
Male	Female
At rest	Moving
Straight	Crooked
Light	Darkness
Good	Bad
Square	Oblong

Table 1: Pythagorean List of Limited and Unlimited Qualities

With the number three, a third element enters into the system and this element adds resolution to conflict and creates harmony. This symbolism is also reflected in astrology by divisions of the circle into three parts of 120 degrees each when planets are in trine. Also the numbers two and three are the first of the even and odd numbers respectively which introduces another Pythagorean principle, the division of all existent elements of the world into the categories of the Limited and the Unlimited (see table 1). Those qualities listed under 'Limited' in some way have the attributes of being more completely defined and tend toward oneness, that is, they either have only one instantiation or exist in a smaller number of varieties. The former criterion can be seen in the pair 'At rest' versus 'Moving' where in an absolutely defined spatial context, 'At rest' is perfectly defined, whereas 'Moving' can be any one of a number of states of motion.[5] The pair One-Plurality demonstrates the second criterion. There are also some pairs which seem to be based on cultural value judgments of the time more than on any sort of philosophical criterion. Among these might be Right-Left and Male-Female. Good-Bad might seem to fall into this category as well, except that Good-Bad as criteria are related by definition to Limited-Unlimited.

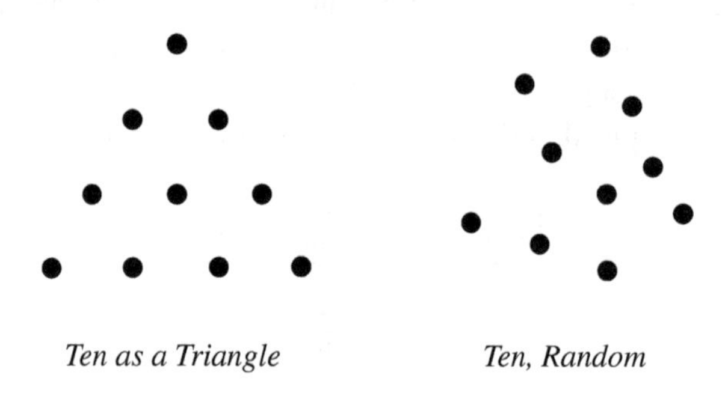

Ten as a Triangle *Ten, Random*

Figure 2

But what is the Odd-Even pair doing here? There are just as many odd numbers as even ones, and, on the face of it, 'even' is no less defined than 'odd.' Here is where another Pythagorean way of looking at numbers, so different from the modern way, again comes into play. Integers are not merely aggregates of units. In addition to philosophical qualities of the kind mentioned previously, integers also have shapes. For example, if one takes numbers which are sums of the following series, 1, 2, 3, 4, 5, 6 … — simply the integers themselves in order —first of all we get one. Then in order we get 1+2=3, 1+2+3=6, 1+2+3+4=10, 1+2+3+4+5=15, so 1, 3, 6, 10, 15. These numbers can be arranged as equilateral triangles. The most famous example of this is the well-known tetractys based on ten. Figure 2 shows ten items arranged in the triangular form and ten objects arranged at random in no particular form. The modern view pays attention only to the number of elements in the set. Both versions of 'ten' are equally 'ten.' But the Pythagoreans clearly saw the triangular form as more essentially 'ten' than the random arrangement. Shapes into which one could arrange the numbers were part of the essence of the number, whereas modern mathematicians concentrate exclusively on the fact that there are ten elements in both diagrams.

The status of odd and even has a similar origin. Instead of having a series which consists of all integers and summing the series at each number of elements, let us make two series, the first consisting of all

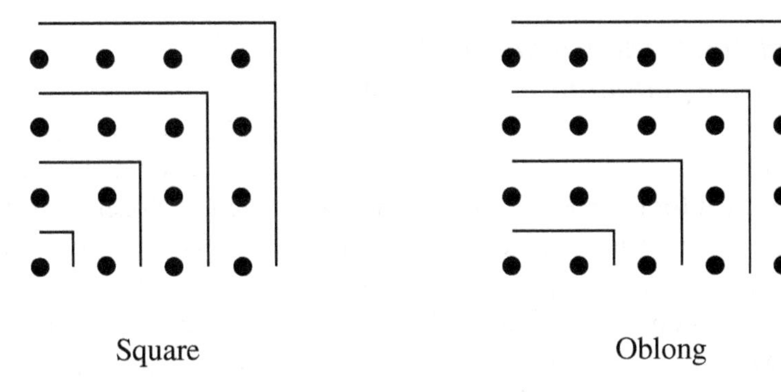

Figure 3

odd, and the second of all even numbers. And, in Square/Oblong in this case, let us start the odd series with one, rather than three (remember that 'One' is on the Limited side of the table along with 'Even'). If we do this, we get the following two series of numbers — odd: 1, 3, 5, 7, 9 …; even: 2, 4, 6, 8, 10 … For sums in case of the odd numbers we get: 1, 4, 9, 16 … These are represented on the left side of Figure 3. For the even numbers we get: 2, 6, 12, 20 … These are represented on the right side of Figure 3. The first set of sums is the numbers familiar to us as 'square' numbers, a representation, which is a survival of the Pythagorean mode of thinking. The second set of sums appears on the surface as a meaningless collection of even numbers. They are, in fact, numbers which make up rectangles when represented geometrically, and can only be represented as rectangles. Furthermore, all squares have exactly the same shape; they may differ in size but not in shape. Therefore, disregarding size, there is only one kind of square. The same may be said for circles. But there are an infinite variety of rectangles, and also ellipses. Thus, the square, being one, belongs on the 'Limited' side of Pythagorean table and the oblong, or rectangle, on the 'Unlimited' side.

There is much more that can be said along these lines, but this will serve our purpose; that is, to demonstrate that the ancients had a very different way of looking at mathematical entities from modern mathematicians, and that their way included not only the most abstract quality of each number, its quantity, but other kinds of qualities, the ability to make certain geometric forms, the possession of philosophical attributes, and so forth. Ancient arithmetic was a science of quality as well as quantity, and the qualities were drawn from visualizable manipulations of the objects which were numbered.

Dimensions: the Cartesian and qualitative 'common sense' views

Now let us turn to the idea of a dimension. Without getting rigorously mathematical for the present, the common sense view of dimensions has its origin first in wanting to know how 'big' a space may be, and second where an object is in a particular area of space. Although the method of representation has been around a good deal longer, the manner in which we define spaces and locate points within them is derived loosely from René Descartes. In this system we take lines along which we will measure, set them at right angles to each other and measure along these three lines in order to define the size of a space or volume, or to locate a point within such a space or volume. For example, we measure a box as so many inches or centimetres wide, another so many inches or centimetres high, and yet another so many inches or centimetres deep. In common mathematical parlance these three 'dimensions' are called x for width, y for height, and z for depth — the well known Cartesian or rectilinear coordinates. The relationship among the various lines along which dimensions are measured is called *orthogonal*, which is derived from the Greek for right-angled, that is, the three dimension-measuring lines, or axes, are at right angles to each other. Figure 4 shows the typical arrangement of the three axes, with respect to each other as well as they can be shown on a two-dimensional page. The z-axis which seems to go down at an angle to the x and y axes must be visualized as coming straight out of the page at the reader.

In ordinary language these three dimensions might be expressed as pairs of directions originating from a central point, usually the observer. The x-coordinate would be Left–Right, the y-coordinate would Up–Down, and the z-coordinate would be Forward–Backward. However,

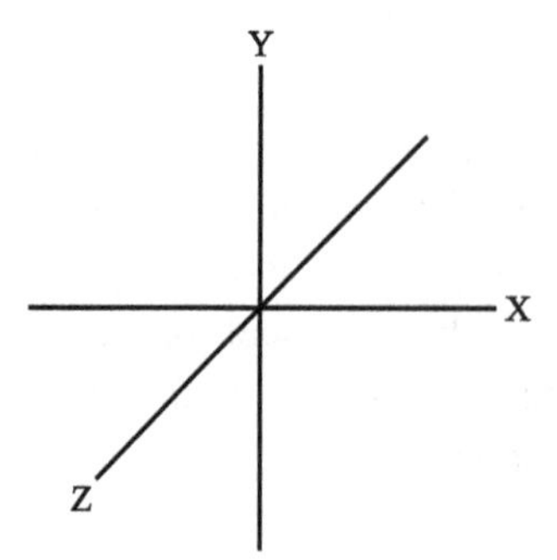

Figure 4: Cartesian Coordinates

unlike the abstract space defined by the x-y-z system, the three ordinary language dimensions have an experienceable aspect and a genuine relationship to the self. The *Sefer Yetzirah*, a text composed somewhere between the first and sixth centuries CE takes these three dimensions a step further toward experienceable and human qualities.[6] A passage from this text, which we will explore in greater detail further on, states:

> A depth of above. A depth of below.
> A depth of east. A depth of west.
> A depth of north. A depth of south.[7]

What we would call distances has been translated in every English version of which I am aware as 'depths.' These can be related even further to human experience in such a way, as we shall see below. This will demonstrate even further that dimensions were not seen by the ancients as mere mathematical abstractions of a purely quantitative or logical nature, but which again, like our integers above, have other kinds of qualities as well. But before we return to this theme, let us look at a line of mathematical logic in the modern (Cartesian) style which illustrates both the power and the weakness of the modern kind of reasoning.

Dimensions, circles, spheres and hyperspaces

In Byrne's edition of *The Elements of Euclid,* the circle is defined in definition fifteen as follows: 'A circle is a plane figure [that is, two-dimensional], bounded by one continued line, called its circumference or periphery; and having a certain point within it, from which all straight lines drawn to its circumference are equal.'[8] Definition sixteen adds: 'This point (from which the equal lines are drawn) is called the centre of the circle.'

However, if we are to define a circle in terms of Cartesian coordinates, it is a set of points which satisfies the following equation:

$$x^2 + y^2 = r^2$$

This relationship is illustrated in Figure 5. In this figure, x and y have the usual meaning of length and height, while r stands for the length of the line going from the centre to the circumference at any particular x and y. The Greek letter phi (φ) indicates the angle between radius connecting the point at x, y to the centre of the circle.

To get a sphere, we merely need to take Euclid's definition of a circle and replace the words 'circle,' 'plane,' and 'line' with 'sphere,' 'solid' and 'surface' respectively. The clear relationship between the circle and the sphere is even clearer using the Cartesian representation.

$$x^3 + y^3 + z^3 = r^3$$

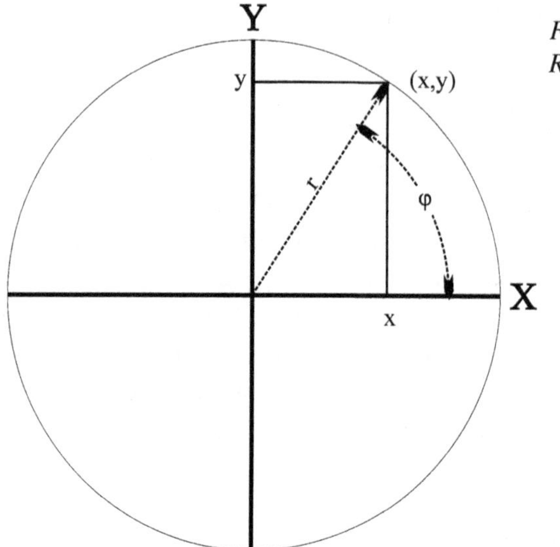

Figure 5: Cartesian Representation of a Circle

The important thing to notice here is that the exponent (the small numbers written above the letters x, y, and z) is the same as the number of dimensions that are needed to define the point. As we have three dimensions — x, y and z — so the exponent must be the number three. Can we continue this process? Yes we can, but for our purposes it is necessary to change our system of notation slightly to continue. If we are to start with a dimensional coordinate called x, and continue through y and z, we will run out of letters. Sometimes authors will use w to label a fourth dimension so that the next equation in our circle-sphere series would look like this.

$$x^4 + y^4 + z^4 + w^4 = r^4$$

However, this still leaves us with the problem of how to notate the next equation in the series. We can solve this problem by designating the orthogonal dimensions not as x, y, and z and so forth, but by calling each orthogonal dimension an x. If we want to represent two dimensions, we use x_1 and x_2 (the subscripts simply mean which coordinate we are referring to). Because we are dealing with complete abstractions here, we can simply refer to x and y as two different instances of a dimension, or x, in general such that the first x is x_1, the second x is x_2 and the third x is x_3 and so on up to x_n in which n stands for the last dimension in use, however many may be required. So here is what we get.

Circle of Two Dimensions $x_1^2+x_2^2 = r^2$
Sphere of Three Dimensions $x_1^3+x_2^3+x_3^3 = r^3$
A 'Hypersphere' of Four Dimensions $x_1^4+x_2^4+x_3^4+x_4^4 = r^4$
A 'Hypersphere' of n-Dimensions $x_1^n+x_2^n+x_3^n+x_4^n+. . .+x_n^n = r^n$

From the point of view of modern mathematics the power of this system is that we can logically and mathematically describe 'hyperspaces' and 'hyperspheres' that we cannot experience or even visualize. Vast quantities of writing in science fiction and physics have been expended in conceptualizing and utilizing 'spaces' of more than three dimensions. The most famous and well-known example of this occurred with Einstein's Theory of Special Relativity, where he found it necessary to postulate time as a fourth dimension and to introduce four-dimensional hyperspace as a 'practical' reality. I place quotation marks around the word 'practical' because the *praxis* in question is that of mathematical physics not ordinary, experienceable reality.

The point I wish to make here is that this is possible only because modern mathematics, since Descartes, has abstracted the idea of dimension from our experiences of right-left, up-down and backwards-forwards, and shorn it completely of the qualities that we may associate with any particular dimension. This is the same kind of abstraction that occurred with whole numbers between the time of the Pythagoreans and modern mathematicians. The fact that we cannot visualize (experience directly) hyperspheres and hyperspaces is not considered relevant, but this raises two questions. Are these 'hyperspaces,' defined by dimension numbers greater than three, meaningfully called 'spaces' in any sense of the term other than an abstract mathematical sense? Is there any possible way that we can experience such 'spaces' as spaces in any visualizable or other manner besides the abstract mathematical? The following experienceable fact opens the problem to question.

The two instances of this set of mathematical and dimensional models that we *can* experience in some manner directly are the plane two-dimensional surface and the solid three-dimensional space.[9] Is there any way in which we can see a three-dimensional space as a kind of *hypersurface* in the same way that we might want to see a hyperspace as a kind of higher-dimensional version of a three-dimensional space? I think that most readers would agree that, aside from the formal, mathematical parallels between surfaces and spaces, the two are quite discontinuous and disjunct. I would argue that we should expect at least the same kind of discontinuity and disjunction between 'spaces' of three dimensions and four, between four and five and so forth.

Many years ago I was trying to explain to a sceptical friend of my father why time could be considered a 'fourth dimension.' He had difficulty accepting time in any way as a fourth instance of the same kind as length, width and height. I said to him that if you were trying to locate an object in space, one would, of course, need to know its x, y and z coordinates with respect to some point of origin. But if the object were moving, one would also need to know when it was there. So in order to completely describe the location of something you would in fact need to have time as a fourth dimension. One would have to have the distances of the object, x, y and z from a point of origin and a dimension t from some temporal point of origin. As plausible as this use of time to locate an object might have been, my father's friend was still sceptical of the idea. He could not see that t=time was a dimension of the same type and x, y and z. I pointed out that I was using x, y, z and t as abstract parameters to locate an object in an abstract 'space' while he was caught on the apparent qualitative difference between the three spatial dimensions and the fourth temporal dimension. I propose, looking back, that he had a point.

The problem here is that I was using time as a dimension in much the same way that I would use any other kind of *parameter*. The *Oxford English Dictionary* gives the following as one definition of parameter: 'An independent variable in terms of which each coordinate of a point of a curve, surface, etc., is expressed, independently of the other coordinates.'[10] Just below it gives another one regarding astronomy: 'Each of a set of numerical quantities, typically six, that jointly specify the orbit of a planet, comet, or other body.'

In the first case, 'parameter' is virtually the same as 'dimension' or dimensional coordinate. In the second, it is simply any mathematical quantity that is necessary to describe something. I submit that in practice the process of abstraction has reached the point where these two uses have become close to identical, that is, that 'dimension' as used in the mathematical modelling practices of Physics is vanishingly close to the same thing as 'parameter,' as used in the second *Oxford English Dictionary* definition. So now in modern string and superstring theory we have eleven or more dimensions. Are these, in any meaningful sense of the word, spaces, aside from the formal, mathematical sense? Using a somewhat absurd example from modern advertizing, let me demonstrate how the collision between the two words, and the conceptual frames they represent, has rendered any real distinction between the two words insignificant.

On the web and on television we have the phenomenon of *eHarmony,* a matching service for persons interested in getting together for relationships. It uses a system of 'dimensions' for establishing relationship compatibility.

> Scientifically Proven Compatibility Matching System: Exhaustive research with thousands of couples found that there are *29 Key Dimensions* of Compatibility necessary for success in a long term relationship. eHarmony is the only relationship site that uses a scientifically proven method to match based on these *29 crucial dimensions.*[11] [italics mine].

Is it possible that Dr Neil Clark Warren and *eHarmony* have surpassed superstring theory, or is it just possible that the word dimension has begun to lose its meaning, and not just in eHarmony? Do relationships exist in a twenty-nine dimensional hyperspace? I submit that the process of abstraction, so beloved of modern mathematics and mathematical physics, has simply created formal entities which can be useful in modelling but which may not be anything more than models, that is, they do not represent exclusively or exhaustively any particular reality.[12]

The Sefer Yetzirah *revisited*

We have previously referred to a quotation from the *Sefer Yetzirah* which discusses the 'depths' of ordinary space. Here is the complete passage which we will now explore further:

> Ten Sefirot of Nothingness — Their measure is ten which have no end.
> A depth of beginning. A depth of end.
> A depth of good. A depth of evil.
> A depth of above. A depth of below.
> A depth of east. A depth of west.
> A depth of north. A depth of south.
> The singular Master God faithful King
> dominates over them all from His holy dwelling until eternity of eternities.[13]

Aryeh Kaplan, the commentator and translator of this work, adds the following commentary:

> Here the *Sefer Yetzirah* defines the five dimensional continuum which is the framework of its system. These five dimensions define ten directions, two opposite directions in each dimension.
>
> The space continuum consists of three dimensions, up-down, north-south, and east-west. This continuum is defined by six directions, and is called 'Universe.' The time continuum consists of two directions, past and future, or beginning and end. This is called 'year.' Finally, there is a moral, spiritual fifth dimension, whose two directions are good and evil. This is called 'soul.'[14]

First of all, the reader should notice that the three ordinary spatial dimensions are not even the first ones listed. Second, note that the *Sefer Yetzirah* groups the three traditional spatial dimensions together, and, according to Kaplan, considers them as aspects of 'universe.' Then we have two other dimensions which refer to 'year' or time, and relationship to God called 'soul.' I have arranged these in Table 2.

Dimensional Aspect	***Sefer Yetzirah* Name**	**Experiental Aspect**
East-West	Universe	Coming-to-be/Passing away
North-South		Dark and cold/Light and warm
Up-Down		Toward Earth/Away from Earth
Time	*Year*	Past/Future
Consciousness	*Soul*	Toward God/Away from God

Table 2: The Experience of Dimensionality in the Sefer Yetzirah.

Here, and previously, I have noted the almost universal tendency to group together the three dimensions of ordinary space as qualitatively different from the usual fourth dimension, time, and any other hyper-dimension. In this table I am attempting to show that one can even make qualitative distinctions among the traditional three dimensions as long as one does not take the already abstracted dimensions of Descartes. We need to take some specific set of three dimensions, not a generalized x, y and z. The *Sefer Yetzirah* does this in the text. Right-Left is replaced by East-West. In many European languages, ancient and modern, the word for east is a word that denotes rising, as in sunrise. According to the *Oxford English Dictionary* the English word 'east' is itself derived from a root meaning dawn, so even our word preserves this sense. Similarly, the word for west in these same languages comes from roots which mean setting, as in sunset. And our word 'west' comes from a root meaning evening. As with dawn, *days* and metaphorically *things* come into being, and with sunset the day ends and things metaphorically pass out of being, so the East-West dimension has the quality of coming-to-be and passing-away, not in time, for time is its own dimension, but with regard to where a thing is in terms of its own growth and maturity. North-South (in the northern hemisphere at least) is an axis of cold and dark versus warm and light. Up-Down has often been experienced as Earth-Heaven, but the presence of Good-Evil as a dimension, if Kaplan's interpretation is correct (and I believe that it is), suggests that the compilers of the *Sefer Yetzirah* did not think of God and Heaven as 'up.' We can make this dimension neutral in this respect by simply stating that it corresponds to being toward and away from the Earth, whatever may be 'up there.' In any case it is clear that the experienced qualities of these three instances of the usual three spatial coordinates are quite different from one another. By making clear, qualitative distinctions among the traditional three dimensions, we have weakened to some degree the perceived qualitative distinction between those three dimensions on one hand, and any additional dimensions we might care to add on the other. All dimensions are different, even though they may share the possibility of being modelled by abstract extensions.

Now we come to eternity. Eternity in Neoplatonic philosophy is a state in which all things that are, ever could be, ever were or ever will be, are simultaneously present.[15] There is no change in eternity. It is described in these writings as if it were a four-dimensional hyperspace in which the fourth dimension, which we experience as change within time, would be perceivable as a linear dimension in a manner similar to the traditional three dimensions. But the *Sefer Yetzirah* describes God as existing in an 'eternity of eternities.' This is an English translation of the same Hebrew phrase that St Jerome translated into Latin as *saecula saeculorum*, literally 'ages of ages,' and usually translated into English as 'forever and ever.' Is this a mere rhetorical phrase denoting a very long time? I think it is not. I believe that the phrase 'Eternity of Eternities' is one level of dimensionality higher than simple Eternity. It is that in which Eternity has existence. Thus if the 'hyperspace' of the *Sefer Yetzirah* is five-dimensional, God is six or more dimensions, if one could possibly model God using a dimensional model. But here is a different way by which we can relate all of the levels of dimension as used here in a coherent set of relationships which are somewhat different from that used in the abstract mathematical model. The following table describes this:

Point	0 dimensions	exists within	lines
Line	1 dimension	exists within	planes
Plane	2 dimensions	exists within	spaces
Space	3 dimensions	exists within	temporal hyper-spaces (time)
Time	4 dimensions	exists within	Soul
Soul	5 dimensions	exists within	God (with possibly inter hyperspaces)

Table 3: The Relationship of the Dimensions

The last of these, Soul, requires a little explanation. First of all, soul here is not the entity often confused with spirit. As used in the *Sefer Yetzirah*, it is a mixture of soul as in world-soul and consciousness in general. The dimension is actually characterized as Good-Evil and, since Evil is the privation of good in Jewish mysticism (as it is in Neoplatonism), and since God is the wholly good (in both systems) going toward the Good is toward God and going toward Evil is away from God. A similar logic permits one to make the claim that this is a dimension of consciousness. God in this system contains all consciousness, and going away from God also presumably makes one go away from consciousness. So how does this relate to our model?

Conclusion: motion in the three levels — Universe, Year and Soul

In the space of the traditional three dimensions (Universe) there can be no motion or change because motion and change of any kind requires Time (Year). Without time there is only each disconnected instant. Time, as we experience it, is the *apparent* flow from one such instant to another. And Time, in turn, requires Soul/Consciousness, because only the awareness of moments just past and the moment now present allows the *apparent* flow of time to proceed. Time and, therefore, the three spatial dimensions, exist within Soul/Consciousness. In this model, without Soul/Consciousness as a dimension in which successive time-spaces exist, it is meaningless to talk about 'objective' reality or existence apart from each individual sentient being or some kind of conscious divinity. And if each individual sentient being is all there is of Soul/Consciousness, if there is no divinity, then I would have to argue that the logic still holds. Whether a rainbow is perceived as an arc of colours above the horizon, or as the refraction of light waves by individual rain drops, whether colour is as experienced by the eye or is represented as wavelengths from the electromagnetic spectrum which are selectively bent, these are events that exist only within consciousness. Consciousness is necessary for direct, sensual perception by the eye, and it is equally necessary to create the model of light as photons, which are sometimes perceived as particles and sometimes as waves. There are no events, no changes, no meaningful definitions of existence outside of consciousness. The only question that remains is whether that consciousness exists only in individual sentients, or exists as well as some form of transcendent being. If the former is true, then we, and beings like us, are responsible for creating the universe! But the experience of mystics through the ages suggests that the latter is true.

Notes

1 Extended magnitudes are those of the kind that could be measured concretely or metaphorically by a yard or meter stick.

2 The word *arithmetic* comes from the Greek word *arithmos* which is the precise equivalent of our word *integer*.

3 In fact, this method of looking at numbers survived well into early modern times. Much of the controversy between the English Rosicrucian, Robert Fludd (1574–1637), and the astronomer, Johannes Kepler (1571–1630), revolved around their two very different ways of looking at number. Fludd came from a completely Pythagorean perspective. Kepler, on the other hand, while he was not completely devoid of Pythagorean tendencies, came from a much more modern perspective.

4 This list is recorded by Aristotle, *Metaphysics* (i 5 986 a 23).

5 Although the English translation is 'at rest' and 'moving,' the Greek words are actually *stasis* and *kinesis*. *Kinesis* is not merely motion in space, but any kind of change, whereas *stasis* is the absence of change.

6 For this dating, see Kaplan, Aryeh, *Sefer Yetzirah: The Book of Creation*, (York Beach, ME: Samuel Weiser, 1990), p. ix. The earliest explicit reference to it dates from the first century, while the oldest quotation from it is from the sixth century. Tradition attributes it to the secret teachings of Abraham.

7 Kaplan, p. 44.

8 Byrne, Oliver, *The Elements of Euclid* (London: William Pickering, 1847).

9 Actually, there is a problem even here. One cannot directly experience a perfect plane surface. All real surfaces have bumps and valleys in them and depart in some measurable way, however small, from perfect flatness. But we can at least visualize a plane in the 'mind's eye' beloved of Platonists and mathematicians.

10 *Oxford English Dictionary*, online version.

11 As stated on their website: http://www.eharmony.com/singles/servlet/about/difference [accessed on 31 December, 2005].

12 By 'exclusively' I mean that this kind of modelling may not be the only way to model, and by 'exhaustively' I mean that this kind of modelling does not exhaust all of the possibilities of the reality being modelled. I am also acutely aware of the philosophical problems of associating any kind of external reality to a model. However, if we take the position that such models are merely instruments created by the mind to make sense of phenomena, then it seems to me to be even more important that we do not neglect aspects of how we experience in favour of mathematical abstractions that we cannot experience. How can we not be able to experience things that we create?

13 All of these quotations are from Kaplan, p. 44.

14 *Ibid.*

15 See especially Plotinus, *Enneads*, V.1.4.

9. Sun Gods and Moon Deities in Africa

JARITA C. HOLBROOK

The belief that celestial bodies such as the Sun and Moon influence and rule the lives of humans is present in several African cultures. The level of regard that people hold for celestial bodies in the selected cases range from believing that they are deities, sentient beings, ancestors, or cultural heroes. All these indicate a belief that these celestial bodies are alive and aware. Their prayers and rituals are one of the indicators of how much they believe that celestial bodies affect their lives, protect them, and keep the world in balance. In some African cultures, the fate of individuals is determined by the position of celestial bodies at their birth. This presentation explores the following two aspects of celestial bodies and humans: 1) that celestial bodies are active and can be appealed to through various forms of worship and 2) that celestial bodies determine one's destiny and are otherwise inactive.

Introduction

Africa has a long and rich history of sky watching, which makes choosing which particular ethnic groups to discuss difficult. For this paper, which was presented at the *Sky and Psyche* conference in June 2005, I chose to systematize the process of selection. Of my library of around five hundred items collected on African cultural astronomy, I have digitized and done a detailed qualitative analysis of thirty articles. These thirty were not selected randomly, but included materials that I was already familiar with (see Figure 1). I targeted the results of selected cultural expressions: astrology and divination, astrology and healing, Sun deity and Moon deity. Many of these cultural expressions appear in one article, for example, an article on the Sandawe of Tanzania has eighteen references to a Moon deity. Thus, focusing on this set of cultural expressions yielded sixteen publications, but only about eight

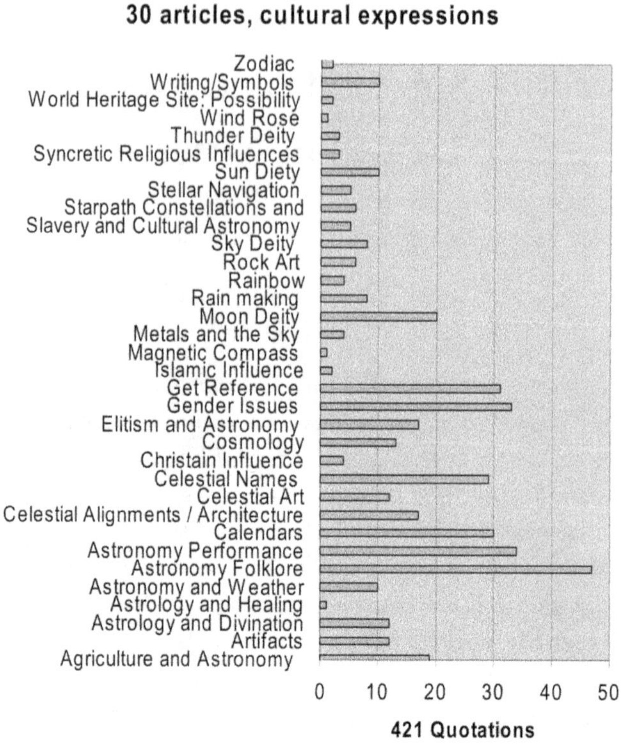

30 articles, cultural expressions

Zodiac
Writing/Symbols
World Heritage Site: Possibility
Wind Rose
Thunder Deity
Syncretic Religious Influences
Sun Diety
Stellar Navigation
Starpath Constellations and
Slavery and Cultural Astronomy
Sky Deity
Rock Art
Rainbow
Rain making
Moon Deity
Metals and the Sky
Magnetic Compass
Islamic Influence
Get Reference
Gender Issues
Elitism and Astronomy
Cosmology
Christain Influence
Celestial Names
Celestial Art
Celestial Alignments / Architecture
Calendars
Astronomy Performance
Astronomy Folklore
Astronomy and Weather
Astrology and Healing
Astrology and Divination
Artifacts
Agriculture and Astronomy

0 10 20 30 40 50

421 Quotations

Figure 1: Results of the detailed qualitative analysis of thirty articles.

unique publications. Looking at the number of cultural expressions and the detail provided in each publication, I narrowed my selection to five ethnic groups: Batammaliba, Benin; Edo, Nigeria; Ngas, Nigeria; Arimi, Tanzania; and Sandawe, Tanzania.

The question of what is and isn't a deity is variable, depending upon criteria determined by the researcher. In order to capture the idea of what constitutes people perceiving celestial bodies as deities, I formulated two questions:

1. Are celestial bodies active, and can they be appealed to through various forms of worship?
2. Can celestial bodies determine one's destiny but otherwise be inactive?

The first question looks at the actions of people to see if they indicate a belief that they can influence celestial bodies. If they pray, make sacrifices, fast or perform other actions then this should indicate their belief that the celestial body is alive enough to hear and see them. The second question parallels many astrological beliefs that celestial bodies are not alive, but they nonetheless influence the lives of humans,and/or determine the fate of humans.

Sun Gods and Moon Deities in Africa

What follows is a brief introduction to each ethnic group, information about the researcher and details about their relationships to the Sun and the Moon, starting with the Batammaliba who live in northern Benin and Togo. They are also called the Batammariba. Suzanne Blier, who is now a professor at Harvard University, is an art historian who lived among the Batammaliba during the 1970s.[1] She focused her studies on the architecture, cosmology and many other cultural aspects of the Tamberma group of Batammaliba.[2] One region of the Batammaliba, Koutammakou, is a World Heritage Site in Togo.

The Tamberma have a complex pantheon that includes celestial bodies. The creator God is *Kouiye* who lives in the west. The Sun is *Kouiye*'s spirit and his creativity. The Tamberma word for the Sun and also 'day' is *Liye*. The Sun is pulled across the sky by two ancestors that are part of the cosmology of the villagers. One ancestor pulls the Sun across during the day, and the other ancestor pulls the Sun underneath the Earth and back to the east during the night. One ancestor is considered red and the other considered black. These two ancestors also influence the layout of the village, on one side of the village being considered the people are associated with the red ancestor, and the other side they are associated with the black ancestor. *Kouiye* is considered the most powerful deity. If rain is delayed during the rainy season, the villagers make offerings to him. The Moon is called *Koutankou*, which means 'elastic.' *Kouiye* created the Moon for two reasons: to be his wife and for timekeeping. The thirteen Moons that make up the year are associated with various tasks that must be done during that month. The months are also associated with particular stars that are celestial deities. The Sun and Moon's children are the stars. When you see 'the Man in the Moon,' it is the Sun spending the night with his wife. The chameleon God, *Koupon*, is associated with the Moon. It is the God of divination. The Tamberma do not make offerings directly to the Moon but to *Koupon*. Thus, the Tamberma, Batammaliba, believe that their actions can influence the Sun and the Moon.

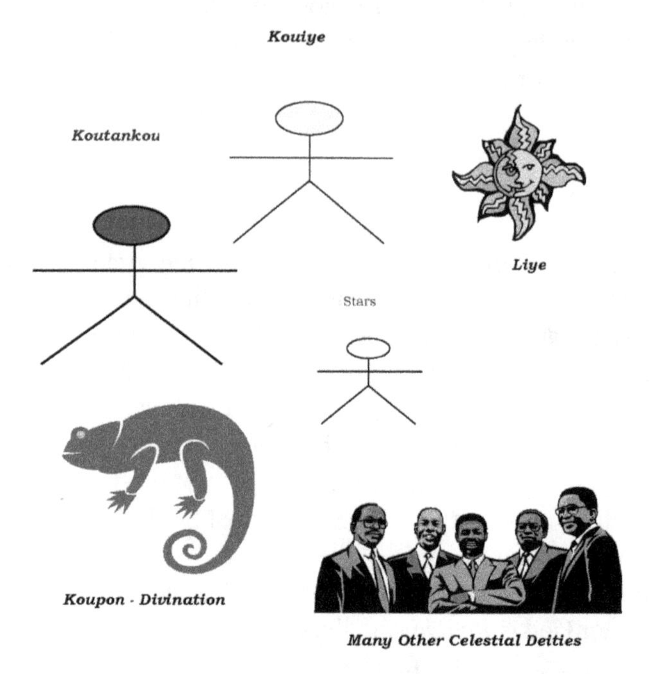

Figure 2: Diagram of Tamberma cosmology. The Sun, Liye, *is a manifestation of the creator,* Kouiye.

The Edo of Nigeria were studied by N. W. Thomas early in the 1900s.[3] Of particular interest is his visit to a shrine in 1910 beside the Ikpoba Road. The shrine included altars to the Sun and Moon. He was told by one of the original priests of the shrine that when the Sun and Moon were fighting, *Osia* and *Olokun* were chained to the shrine until they resolved the quarrel. Thomas thought that the priest was referring to an eclipse, but couldn't get confirmation. Who were these two individuals that were chained to the shrine? *Olokun* is a deity of the deep waters among the Yoruba people who live in western Nigeria.

But *Osia* is unknown. It is unclear if these were simply young people associated with the shrine, or if they represented the Sun and Moon in some ceremonial way. There simply is not enough information presented. Nonetheless, what is clear is that the Edo believed that their actions could influence those of the Sun and Moon. Another part of the shrine included a room with a pond, this was the site of some of the rituals. When appealing to the Moon, they would draw the sign of the Moon in chalk, next to the pond. For the Sun, they would make a chart mark, but also leave an offering of Cam wood. Thomas also recorded sayings. During the new Moon, the Edo would throw sand towards the Moon and say 'if you are a good Moon, bless me; let me reckon you as a good Moon for my "good luck"; if you are a bad Moon, I run away.' Similarly, chalk would be thrown, and they would say, 'I give it to you, do well for me.' Two other sayings dedicated to the Moon were: 'Moon, now I see you, let me be well and my son too,' 'I see you go, let me get money.' The symbols, offerings and sayings all indicate that the Edo believed that their actions influence the Sun and Moon.

The Ngas of Nigeria live on the Jos Plateau. In the 1970s, Deirdre La Pin was part of a film crew documenting the life of the Ngas.[4] She did a follow-up study in the 1980s. She recorded that the Ngas say that the Moon is not a God. However, here I present parts of La Pin's research that may indicate otherwise. The biggest ceremony, among the Ngas, marks the end of the year in the beginning of the next. It is called Mos Tar — the beer of the Moon. The activities associated with the ceremony are presided over by a priest-astronomer called the Golong. The ceremony begins seven days before the next new crescent Moon is sighted. The Golong has to time everything exactly. If the week-long ceremonies do not result in a sighting of the first crescent, everybody would get sick.

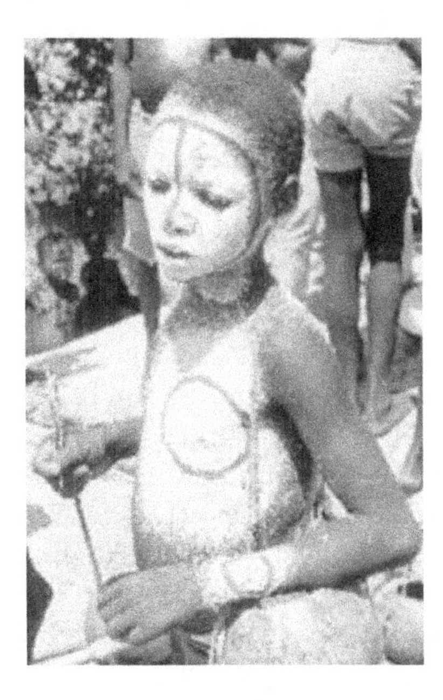

Figure 3: A Ngas 'warrior' ready to 'Shoot the Moon.'

The evening before the first crescent is sighted, newly initiated boys must 'shoot the Moon.' Their actions are to hunt the old Moon, shoot it and bury it, allowing the new Moon to appear for the new year. Similar to the Tamberma, the Ngas name their months according to the activities that must be performed during that month. For example, they have the Moon of harvesting. La Pin recorded that the Moon is considered to be a self-willed being and not a deity. Yet, the Moon is a source of all life and it regulates fertility. The Moon makes time. Moonrises on the horizon mark the days of the Ngas calendar and are referred to by pointing. The Moon is both male and female because it is both nurturing and associated with fertility, but is also a hostile enemy. During eclipses, the people shout 'the Lord is dead' and beat the funeral drums until it is over. During Mos Tar, the people sacrifice a ram to the Moon. During the final Moon shot, the young warriors shout 'Ho, where were you when we shot your brother?' This refers to the fact that they shot the Moon the year before and just shot it again to initiate the New Year. Clearly, the Ngas believed that their actions influence the Moon, and the Moon influences their lives.

The Arimi of Tanzania are also known by their Swahili name, *Wanyaturu*. The following is based on field research conducted in the 1960s and 1980s by two different groups.[5] There are three important celestial bodies in Arimi cosmology: the Sun, the Moon and the Pleiades. These three are called the lamps of the Gods. The Sun is relatively unchanging, though the position of the Sun changes on the horizon over the course of the year and it has Sunspots, but compared to the Pleiades and the Moon it is considered to be unchanging. The Moon in contrast, changes shape and position over the course of a month. Both the Moon and the Pleiades are not always visible in the night sky, thus they are said to be subsumed by the Sun. In Arimi cosmology, the Sky God, *Marungu*, is the most important and presides over all things. A triad of former ancestors that are deified are *Matunda* the Creator, *Magheme* the thinker and *Mahyana* the speaker. These three made all the things on the Earth with the approval of the Sky God. The Sun is called *Yuva* and the Moon *Mweri*. Each morning the sunrise is greeted and praised. The Priest, who is also a diviner, praises the Sun before every ritual as well.

Over the course of the day, the Sun changes characteristics: in the morning the Sun is the diviner and healer. People appeal to the Sun to optimize those qualities in their lives. The midday Sun is the most powerful. It is associated with fertility, yet its heat can be deadly. The afternoon Sun is *Musimbuya* — the finder of sorcery, it absorbs all evil.

At sunset the Sun and evil are buried. The Sun is said to enter its purifying bath.

Over the course of the month, the Moon is associated with different characteristics. At the first crescent, the Moon is said to be male and it is the time of conception. At full Moon, the Moon is said to be female, it is an auspicious time for ceremonies. When the Moon is said to be female, it is the time of Earth and menstruation. An Arimi saying when women see the first crescent is: 'O Moon; take away all our evils, troubles, bad thoughts, sufferings and children's diseases. Men, go into your wives. They all women conceive well.'

The Sun and Moon are part of the of the Arimi pantheon. The Sun and Moon are both praised and appealed to. They are seen to influence the lives of humans, but humans can also influence them.

The Sandawe are neighbours to the Arimi in Tanzania. In the 1960s, Eric Ten Raa did a detailed study of the Sandawe and their

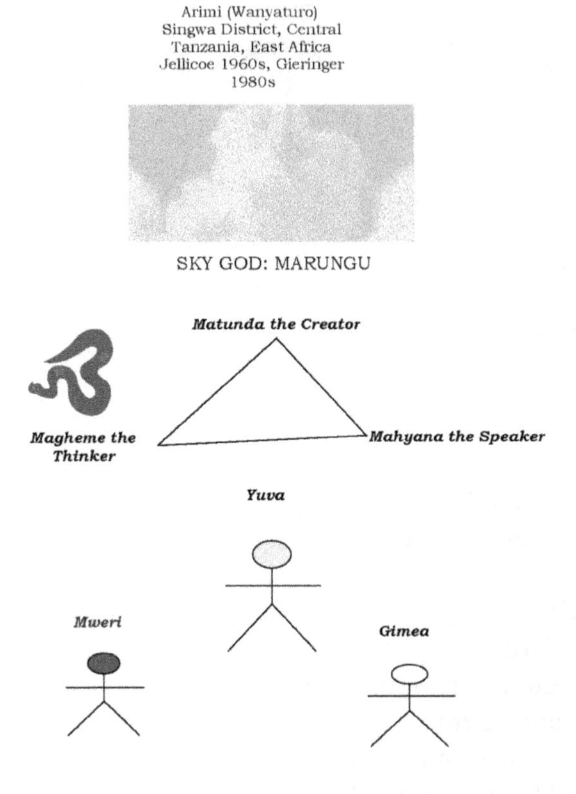

Figure 4: Diagram of the Arimi Cosmology

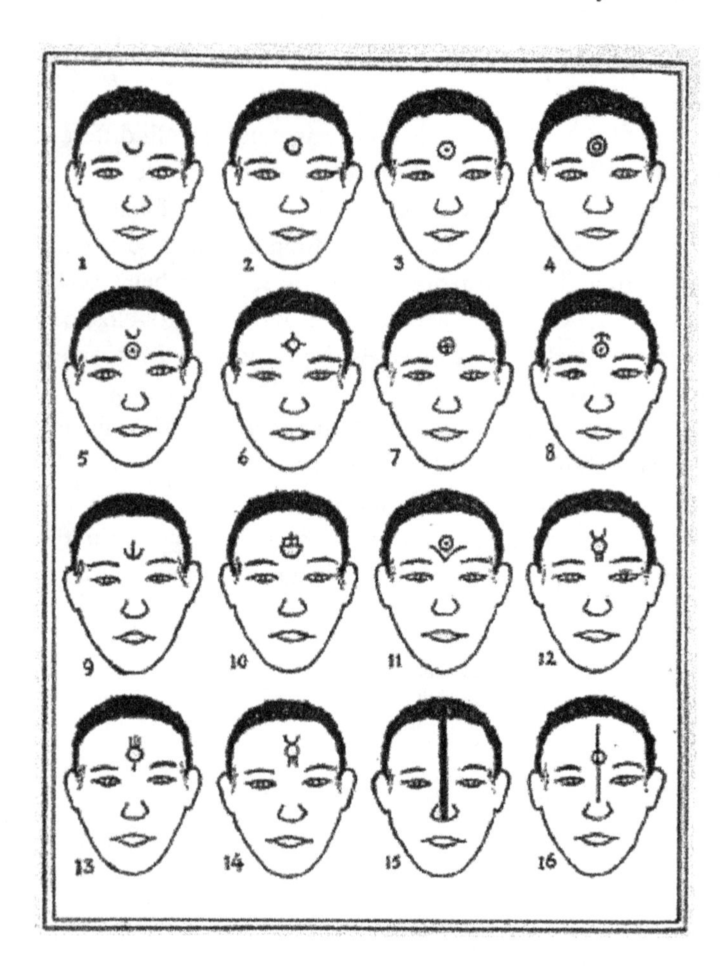

Figure 5: The Moon is present in ten of the sixteen facial marks of the Sandawe, collected by Ten Raa.

relationship to the Moon.[6] Among the Sandawe, the Moon is considered to be a supreme deity associated with fertility, coolness, rain and beauty. The Moon is considered to be female and beautiful. The Moon is believed to preside over the gender of babies: if conception takes place during the waxing Moon, it will be a girl. If conception takes place during the waning Moon, it will be a boy. After death, the last mourning rituals have to also conform to this standard in that if the dead person is female, the last rites must take place during the waxing Moon, if male, during the waning Moon. The Sandawe honour the Moon during the full Moon with a fertility dance. After giving birth, the new mothers show their baby to the new Moon and do a swinging the baby ceremony in the moonlight. The crescent Moon resembles horns, and horned animals are associated with the Moon. If a halo is seen around the Moon, they know that it is going to rain soon.

In contrast, the Sandawe believe the Sun is male, but deadly. It scorches the Earth and causes fire. They believe that at first the sky was ruled by the Moon and everything was cool and there was plenty of water and the land was fertile. The Sun lived with humans during that time but fell in love with the Moon. So, he became the Sun and took the Moon as his wife. The Sun rules death. If a person dies at night, he must be buried before the next Sunset. If he dies during the day, he must be buried the next day. The Sun leads the souls to the west, so the dead must be laid to rest with enough time to follow the Sun westward.

The Sandawe have rituals and dances dedicated to the Sun and Moon but they do not make sacrifices to them. Ten Raa considers this to be an indication that the Sandawe do not consider them as deities. However, the Sandawe believe that the Sun and Moon influence the lives of humans, but humans cannot influence the Sun and Moon. Rather, if something is going wrong, ancestors and other spirits are blamed and sacrifices and appeals are made to them rather than the Sun and Moon.

Conclusions

These five ethnic groups present an interesting cross-section of the relationships between Africans and celestial bodies. The examples presented were chosen because there was some indication that the Sun and the Moon were deified in their cultures. However, the definition of what is a deity and what signs indicate that people consider something to be a deity are fluid. Thus, when doing field research in Africa, the researcher and the people may have conflicting definitions. For this work, my focus was on the area of influence as an indicator of deification to some extent. I looked at the actions, especially rituals, ceremonies, and sayings directed towards the Sun and the Moon. To me, these indicated that humans could influence these celestial bodies. I then searched for the reciprocal: indications that people believed the celestial bodies influenced their lives.

Ethnic Group	Sun	Moon	Prayers/ Appeals	Shrines/ Offerings
Batammaliba	(God)	Goddess	No?	Yes
Edo	n/a	n/a	Yes	Yes
Ngas	n/a	Personified	Yes	Yes
Arimi	God/Goddess	Goddess	Yes	n/a
Sandawe	God/Goddess	Personified	Yes (rituals)	No

Table 1: Sun Gods and Moon Deities.

Ethic Group	Sun / Moon	Influences Humans	Humans Influence
Batammaliba	Sun	Yes	Yes
	Moon	Yes	Yes
Edo	Sun	n/a	Yes
	Moon	n/a	Yes
Ngas	Sun	n/a	n/a
	Moon	Yes	Yes
Arimi	Sun	Yes	Yes
	Moon	Yes	Yes
Sandawe	Sun	Yes	No
	Moon	Yes	No

Table 2: Summarizing Table of Influences

Each of the five ethnic groups in this sample has some degree of deification of the Sun and the Moon under these criteria. Tables 1 and 2 summarize the details for each group. In Table 1, if an explicit statement has been recorded saying that a celestial body is not a God, it is reflected in columns one and two. If there is some question about their beliefs, there are parentheses. Otherwise, I use the fact that appeals are made as an indication of deification. Generalizing for these ethnic groups:

1. The Sun and Moon can be influenced by the praises, rituals and offerings of humans.
2. The Sun and Moon influence the lives of humans.
3. However, the Sun and Moon are not always Deities, but can be supreme personified beings, former humans (Sandawe) or a tool of the Gods.

A final note: the pantheons of these ethnic groups often extend beyond just the Sun and Moon, and the other Deities are not presented in this work for the most part. However, details of the remaining Deities can also be found in the referenced materials.

Future research

Of what significance is the deification of celestial bodies to contemporary anthropological studies of Africans? An over-generalization is that Africans live in a highly variable and stressful world given civil unrest, the processes of modernization, the introduction of new religions, epidemics, droughts and famines. Uncertainty and risk can be managed in a variety of ways; one is to use the services of psychics or diviners more frequently, another is to increase religious practices mainly in the form of prayers and appeals to the Gods. Studies such as this one serve many purposes for future anthropological research starting with comparing practices today with those presented here as a way to track changes in religious beliefs over time. The link between environmental adaptation and the deification of celestial bodies is not well studied and begs the question whether this leads to more sustainable practices and less environmental degradation. This study provides a snapshot of African peoples' perceptions of the two major celestial bodies and their importance for their health, agricultural activities and their souls.

Acknowledgments

I would like to acknowledge the Sophia Centre staff for providing the opportunity to present this paper and funds to support my stay in Bath during the Sky and the Psyche conference. I would like to acknowledge the financial support of the University of Arizona for my travel to the United Kingdom.

Notes

1. Blier, S. P., *The Anatomy of Architecture: Ontology and Metaphor in Batammaliba Architectural Expression* (New York: Cambridge University Press, 1987).
2. As above, as well as Blier, S. P., *Stargazing in Tamberma: Ceremonial Structure and Symbolism* (Unpublished), and Blier, S. P., *The Cosmological Dimensions of Tamberma Architecture* (Unpublished, *c.* 1981).
3. Thomas, N. W., 'Nigerian Notes, IV: Astronomy,' *Man*, 19 (1919), pp. 179–183.
4. LaPin, D. A., *Sons of the Moon* [videorecording]: a film by Deidre LaPin and Francis Speed., located in Ngas, Central Nigeria, F. Speed, Editor (Berkeley, California: University of California Extension Media Center, 1984).
5. Jellicoe, M., P. Puja, and J. Sombi, 'Praising the Sun,' *Transition*, 6:31 (1967), pp. 27–31, and Gieringer, F., 'To Praise the Sun — a Traditional Nyaturu Hymn of Praising God and Asking for His Blessings,' *Anthropos*, 85:4–6 (1990), pp. 518–523.
6. Raa, E. T., 'The Moon as a Symbol of Life and Fertility in Sandawe Thought,' *Africa*, 39 (1969), pp. 24–53.

10. Where the Heavens Meet the Earth: Inspirations from the Lives of Carl Jung, Jalal-u-din Rumi and Mahatma Ghandi

NICHOLAS PEARSON

This paper is a personal odyssey using the insights, inspirations and teachings of these three great men to examine where heaven might have gone. The work of Carl Jung is explored alongside the Sufi world of Beauty and Vision, and how these deep inspirations affected the author. The paper then explores Ghandi's key ideas stemming from his readings of the Bhagavad-Gita *and the centrality of the* Gita *to our own modern condition. The practical impact of this work in psychotherapy is discussed.*

In this paper I will try to take you off in search of inspiration by dipping in some ancient springs where we may hope to find her. I have chosen to imagine that we might find her at the place where two opposite sorts of worlds meet, where the Heavens meet the Earth, for example, or perhaps between the visible and invisible, or even between man and woman. I am drawn to this subject of inspiration because, in an increasingly grey, secular and chronically unstable world, it may well be the place where we can pick up the end of a thread that just may re-orientate us towards a world we have lost, a world that our forebears used to call sacred. So, let us see if together we can find some inspiration from these ideas.

I am mindful that today, in our increasingly one-sided rationalizing world, those who actually think or have any real intuition of a world we might call sacred are becoming a smaller and smaller minority. I really do not think we have begun to understand how catastrophic this is. There is, of course, a huge increase in all sorts of therapies and spiritual fashions, but most of them put man as the great manipulator of his own destiny and they never really accept the sense of a presence in our lives, another root in our lives, upon which our lives depend just as much as they do upon the skill and cleverness of the rational mind.

To be clear, I have nothing against the rational mind, without which we would be in a truly sorry state, but any totally one-sided view of life is, as Jung himself says, simply barbaric. It's as if we need two opposing principles to reflect reality, one just won't do. This is an old story, but one that has now become acute. We see it starting as a tendency of different ways of looking at the world and that tendency manifests itself in the thought of the great spiritual philosophers of our past — Plato and Aristotle, Aquinas and Eckhart, Averroes and Ibn Arabi — and we may be sure that difference is with us today in the different ways you view your different worlds and even in the different ways you may read this chapter. You can call it by many names, but for the moment let's just call it the difference between Faith and Reason, a tendency that grew with the centuries into what has now become a real and damaging split.

In my rather wandering journey I have chosen Jung, Rumi and Ghandi as reference points because not one of these really great men were just observers, or just pure thinkers or pure intellectuals. All of them put their own lives into the crucible of their own lived experience of, and in, the world. What they learned and, in the end, what they taught was that we live in a world of which we are not the ultimate masters. It is for this reason that I find their lives a cause for inspiration, particularly now when the world shakes with uncertainty and fear, and faces a future it is not sure it can control or even understand.

It is the ideas of these great men that inspired me, so our journey with them will inevitably be something of a personal odyssey, a personal journey. And while I don't pretend to be able to tell you exactly where heaven has gone, it seems to feel increasingly as if it sure ain't here. My own intuition tells me that inspiration may be one of the ways to pick up the end of that golden thread that points us again in the right direction.

I have chosen Jung as my first stop because it was finding Jung at a particularly turbulent period of my own life that set the course and direction of my own search. I started to search, for the simple reason that I had to. I had stood for Parliament at a young age, thinking from my upbringing and total lack of real qualifications that I was most perfectly suited to save the whole western world single-handed; arrogance and an unreasonable and unreflected self-confidence being just two of the compensations we can take up to hide feelings of uncertainty and vulnerability. Not an observation, you will understand, that the patriarchal world or the old tribal cannon wish to hear much about. Well, to cut a long story, my path to certain glory was rather violently terminated by a horrific accident in which I killed a man while driving my car into work one morning. I lost my licence, my seat and the whole purpose of my life that I had fantasized I had been born to fulfil.

I remember sitting in the wreckage of the crash, dumb with horror and sorrow and finding that, as I waited for the police, a thought kept coming into my mind: 'now I must go to a monastery.' That was a metaphor, as it turned out, but an accurate one in that I needed most desperately to take stock of the whole direction of my life and the creation of a monastic-like space was exactly what I needed in which to reflect. In time I found this space in Jung.

I discovered Jung in a dream; very Jungian some might say. I had gone to the Sun, to Italy, to recover from the complete emotional exhaustion that had followed my crash and the destruction of my world. While I was there, I dreamed that I was sitting on a beach and holding in my hand a plant that was attached to the soil by a thin root which was about to break. This summed up pretty well my emotional and mental state at the time. A voice from all around spoke in my ear the name of Jung and, as if commanded, I placed the plant in the soil and it started to grow. Well you don't have to be Freud to work that one out, save at the time I knew nothing much about the great Swiss psychologist Carl Jung. As a result of the dream, I bought twelve volumes of Jung's collected works and sat alone in my cottage one winter and read the lot.

Put simply, Jung and his whole body of work gave me the courage to begin to explore my own inner world on my terms. My whole previous orientation had been 'out there,' it seemed that my whole education and the culture from which I came concerned itself principally with everything 'out there.' And out there was ruled by some pretty stern injunctions as to who I was and what I could be and do.

Jung himself had faithfully followed that great injunction 'physician heal thyself,' one that as a psychotherapist I seek to remind myself of frequently. He had done this by turning inward and honouring, with all the faithfulness of a priest, the images arising in him. It has always puzzled me that those who revile the art of psychotherapy — and there are many — are often perfectly happy to accept the Christian injunction that the kingdom of heaven lies within, and yet, when asked to actually examine what 'within' might mean in the context of their own lives, their own interior lives, they shrivel up with contempt, or rage or fear, which is, I suppose, much the same thing. They then accuse one of introversion, indulgence and selfishness and sail majestically and certainly on their way. But if we are to seek the kingdom of heaven and if, as we are told by almost all the great traditions, it is within, how to start if not with one's own self?

Burying myself in Jung taught me, above all, the real value of the imagination, a truly vital tool without which we will not get far on this path of investigation, or any other spiritual path come to that. This active imagination of Jung's had of course nothing to do with what we might call 'imagination.' This imagination is the faculty *par excellence* that has the capacity to build bridges between worlds, between Sky and Psyche, Visible and Invisible, Man and God. It is, in my experience, *the* primary tool through which the place of inspiration, meaning, and faith in the world, is reached.

Jung himself, deeply concerned by the collapse of Faith in the face of the overpowering legions of the rational literalizing mind, has this to say about our current faithless state. 'The evil we are suffering from is less a split between Faith and Reason than between Reason and an Imagination that has become incapable of establishing an accord between the two parts of the Universe — the visible and the invisible.'

Today our established mindset is overpoweringly rational. What we cannot weigh or measure does not, cannot, exist. This is an impoverished outlook, for it is one-sided and, therefore, unbalanced. The imagination has become largely degraded to that which is now considered to be largely imaginary. The whole area of religion has suffered a total and stunningly swift collapse through this degradation and misconception. Anything to do with a God has become simply imaginary. But without imagination one cannot — I most certainly could not — see or hear divine things. I could not place my own life in a context that felt meaningful beyond that of surviving, getting and spending. The Sufis of Islam say we in the West have developed an 'Industrial Consciousness' which brings huge material benefits, but is wholly incapable of any real deep spiritual insight. We try so hard to do it with the intellect alone and we wonder why we end up chasing our tails; we no longer know how to enter unseen worlds. The world I had lived in honoured this invisible world only to the extent that it supported the secular aims of the country. I think this is still true. So we were split, 'slightly schizoid,' saying one thing and doing another.

Jung saw, and in time I came to see, just how enshrined that dominating, demanding God of the Old Testament had become in the whole way we functioned. Masculine, aggressive, heroic Reason let loose without the mitigating 'slowing down' aspect of Feminine Soul, split off from the whole, and because unrestrained, became all-powerful, for it is no longer balanced by its opposite. Self-centred humanity, driven principally by fear, takes the god-like powers upon itself, until any real notion of Source, of Origin or of a Thou is forgotten, because by now, in our one-sided Promethean certainty, humanity thinks it does not need 'the other side.' Indeed, the ambivalence that this would cause would hinder the unshakeable certainty that we humans are the sum of all things, that we can do what we like. And yet, a terrible fear lurks underneath, a terrible uncertainty. Our world trembles with it.

The journey I undertook in my analysis began to teach me just to what extent my inner world bore on my outer, and this insight led to what I came to consider as Jung's greatest gift to me. It's a subject we shall pick up again in the Sufi tradition and one that is, I believe, the only route back to a more harmonious world, more harmonious because more balanced.

The world of the feminine

I use gender distinctions here with caution because the lines become blurred. I am referring here more to qualities and shades rather than actual gender distinctions. The Feminine, the Anima, the Goddess, or, to bring it down to earth, my own relationship with what we might call the gentler virtues — reflectiveness, mercy, compassion and women themselves, without whose capacity to hold and mirror and to reflect the reality of Beauty (and here I will use my gender bias), where indeed would we poor males be? I am not sure how many women would agree with me when I say that misogyny, or perhaps fear of the feminine, is still today a widely suffered condition of the male, a misogyny that seems to go with a certain hardness of heart.

I have to admit that at that time I was no exception to this disturbing condition; a condition that I feel is central to the resolution of our quest. Some progress has indeed been made in how our culture relates to the feminine and to feminine virtues, and I am, of course, talking not only of men to women, but most particularly to those same feminine virtues within man himself and woman herself. But we must understand that, for all the progress, unconsciously, many of the old inherited derogatory structures are in place and still profoundly affect the way we operate as individuals and as a culture. We see these structures in the invasion of Iraq, the wanton destruction of so much beauty in the lived and built environment and in Nature, the aggressive obsession with this wasteful, dangerous and unsustainable profligacy we call the economy and the paranoid obsession with 'the enemy out there.' Jung taught me that Inwardness is a Feminine quality, so any Seeker is going to discover, as I had to do, that it is through the primary feminine symbols that the road turns inward towards reflection and eventually, blessedly so, to a more balanced state of Being as well as Doing.

I came to see that so much of my rush, so much of my general anxiety, so much of my feeling that if I did not hold the sky up, who would? So many of my very twentieth-century angsts were due to a lack of that essential feminine quality of Reflection and Repose. This turning to the Feminine, turning inward, was the path I had to take in the hope that, as the Song of Solomon says of Wisdom, 'In the end she will return us to the straight road and reveal her secrets to us.'

This issue of the feminine was the single biggest reality (apart from mere terror) that I had to deal with in my own adventure of discovery. Let me try and show you a little of how it worked in me and how I came to see, through working with Jung's tools, in what unconscious contempt I held that side of my nature and the world. I had a dream which I interpreted as revealing the sick and neglected woman within. In reflecting upon this horrifying image, I turned inwards and as I paid attention to this area of my life, as I reflected upon this enslaved, sick, neglected woman within and what she might mean, my interior world began to change. But as, to use a metaphor, the footsteps of the Goddess approached, as the feminine world, the feminine side of the psyche, reasserted itself in my psyche, the anxiety in my life increased to fever pitch until my whole being was awash with terror as my conscious mind shrank before a power that was quite clearly infinitely greater than itself. That, however, is really a story in itself, suffice to say that after a tidal wave of terror had washed over me, I began at last to enter a new, more secure and more meaningful, world.

Jung has, like any great man, detractors, but what no one can take from either him or Freud is their rediscovery of the reality of a living interior life and its reintroduction to the cultures of the west. It is around this and the issue of the Feminine with which our world is going to have to come to terms. She will return us to the straight road and reveal Her secrets to us. I feel sure that we are going to need those secrets for our very survival

Our second watering hole on this journey is with the Persian Poet Jallaludin Rumi. If my journey with Jung restored to me a deep sense of the interior world and its treasury, Rumi gave me back a sense of Beauty and reinforced in me the vital practice of working with internal imagery that I had discovered in Jung. The Sufis are the Gnostic arm of Islam, often in conflict with the more dogmatic side of the tradition for their liberal ideas and their desire for a direct relationship with the divine. They widely and generously accept other traditions of faith and openly declare that a man might need all of those to come to the presence of God:

> Earnest for truth, I thought on all the religions
> They are I found, one root with many a branch
> Therefore impose on no man … a religion
> Lest it should bar him from the firm set root
> Wherein all heights and meanings are made clear
> For him to grasp[1]

The first Sufis were the Mevlavi Dervishes, started by Rumi in 1207. But from the thirteenth century onwards, flowered an astonishing succession of philosophers and poets who can teach us things about our intrinsic natures we have not even dreamed of; philosophers such as Ibn Arabi, Avicenna, Suhrawadi, and poets such as Hafez, Attar, Mirabai and Kabir. To reach the rank of Philosopher in ancient Iran was to have come to the place of ecstasy in yourself. How do we square that with our so-called modern philosophers whose untethered mental speculations have led, I believe, to much of our modern confusion and to our loss of the sense of the Divine in our lives and culture. Sufi dance and meditation led to a direct experience of being filled with Divine Ecstasy, their practice and language is full of it. See it for yourself and it may provide you with a pointer to where the sacred might have gone and where it still most surely is. Why did Rumi's poetry touch me so deeply? Well firstly there is the sheer beauty of it, and Beauty is at the heart of the Sufi idea of God. 'God is a beautiful presence longing to be known' and his presence manifests itself through Beauty, in people, objects, values, through Longing and through Joy.

But not only is there a staggering beauty and truth in Rumi's poetry, but his prose and the insights that flow from it are so clear, undogmatic, and psychologically astute that, to my mind, they can only flow from a direct relationship with a world that has properties we must surely call sacred, and that we seem to have lost. Let's look at some of them. A man comes to Rumi and says 'Please God that I could go to the other world, there at least I could be at peace in the presence of my Creator.' Rumi replies: 'What do you know of where He is? Everything in all the worlds is inside you. Whatever it is that you are hungering for, work for it here, by yourself, for you are the microcosm.'[2]

I think some of the best of modern psychology supports this. I was taught 'The world is never as you think it is but it is as you are.' And this is so for the simple reason, as Rumi says, 'For you are the microcosm.' Through this insight I began to see how my inner world really was connected to my outer. I first noticed this connection of an inner and outer Beauty through a dream in which a beautiful bird, a conjunction between a hawk and a dove, came and sat on my arm and talked with me. What struck me was how, from that time on, I only had to see an actual hawk wheeling in the sky for my heart to tumble with the connection and to afford the real hawk in the sky an almost palpable sense of reverence and awe. Part of my inner world had connected to part of my outer and the result was the first intimations of what a sacred world might feel like and mean — that is, to be truly at home in the world, to be fully connected to it on all levels.

I think the best way for us to go deeper into this world of personal vision is through one of the great injunctions of mainstream Islam. We will then look at what we mean by the word *Theophany*, which is central to Rumi's whole idea of the Beloved, occurs in his poetry again and again and is central to our need to find this land again, to find 'that which is Beloved.'

In the Koran it is written: 'Whereso'ere you turn your face, there is the face of God.' An intuition that the world on some level is a form of manifestation of the Divine. As Rumi says, all reality is Theophany. There are no autonomous things, only manifestations of the breath of God bound together in the great community of Beings that is the Creation.

Well, the rational mind can tear that to pieces in a second and say 'yes fine, but what do you actually mean by the 'Breath of God,' where is it, how can you prove it?' But the heart in the Sufi tradition is the real organ of perception. The heart, the poetic imagination, will know there is a truth here, it can attain the knowledge that this is true on some level, and will be inspired as a result *in spiritu* — the Spirit will move inside in assent, in recognition of a truth

The Sufi experience, Rumi's experience, centred on this insight of Theophany, of being aware of a Presence. The presence of God, in whatever form it has chosen to reveal itself, rises up in front of the observer and presents itself, announces itself to the observing mind and eye and, of course, to that mighty builder of bridges, The Holy Imagination. So, to use our own tradition as an example, Moses sees the Bush is burning, when someone without this organ of perception, the imagination, may just see a bush.

Now, the Sufis perceived an intermediate world that lay at a different level of consciousness from the gross material world in which we normally live and which could only be bridged or entered into through the imagination. This intermediate world, the *alam al mithal* as they called it, was the place where Spirits became embodied and Bodies became spiritualized. This intermediate world was the ground of Vision and of Prophecy. To enter it required the Imagination. It is, I am sure, why our great English mystic William Blake granted the Imagination such a place of primacy in his visionary way of seeing things so as to call it 'Jesus the imagination,' for imagination is the only tool to enter that intermediate world of vision and prophecy, a world the Sufis would claim to be more real than the one we consider material. Through the bridge of the imagination, an unseen Holy world can become manifest. Moses' perfectly ordinary bush begins to Burn with Holy Fire. God, in one form of his creation, presents himself to the perceiver in a Theophany; burning bush, beloved, voice, event, angel of annunciation. This is not historical myth buried in the mists of history, it can happen and does happen today, including in good psychotherapeutic practice which is grounded in the world of interior transpersonal images.

Of all the Theophanies of the Sufi Tradition, perhaps the greatest and most treasured are Beauty and the Divine Feminine. You will hear in the Sufi experience of their quest for the sacred echoes of the Fedeli D'amore of Dante, and the whole chivalrous attempt by the Troubadours of Mediaeval Europe to reconcile spiritual and physical love. The Sufis realized that a purely spiritual love was only half the picture, and a wholly physical love was only half the picture, but only by going through both might a third possibility become manifest. In the case of Rumi this process was ignited by his relationship with his spiritual mentor, Shams of Tabriz, but it led through this concrete earthly fact towards the final mark of his quest, the love for his own Soul, his own Divine centre, or to use his own language, the search for His own Beloved:

> How can you ever hope to know the Beloved?
> Without in every cell becoming the Lover
> And when you are the Lover at last, you don't care
> Whatever you know or don't know, for only Love is real.[3]

This inward journey commences with the injunction similar in kind to that of the Greek oracular tradition at Delphi; similar to that which we espouse as psychotherapists: 'He who knows himself, knows his Lord,' a great invitation to the interior journey. The second great Sufi injunction was to 'Worship your Lord as if you could see him.' Again, this is not going to happen without the Imagination, or, indeed, without the poetic eye of the heart. But when it does, as it can, when deep in the attention of an inward vision this perception is achieved, it leads to this astonishing and beautiful insight:

> I saw my Lord with the eye of my heart
> I said to him 'Who art thou.' He answered 'Thou.'[4]

Has there ever been a clearer announcement of the presence of the Divine deep within the human heart? And what, you might be tempted to ask, has all this to do with twentieth-century man mortgaged up to the hilt in so many ways? I can answer you and say everything. Good psychotherapy is just such a journey to discover these interior images and, if undertaken with faithfulness, will lead to the discovery that these interior images are still there for us, still there for our help and guidance and our inspiration on our long night sea journey, which is so much part of life. The feeling experience that comes from working in this way is that these images do contain energy, which has the power to transform the conscious mind. To begin with the imagery of Rumi's work may be appropriate to our lost state. How badly the gift of these interior images needs our attention and care, and how we will be rewarded if we give it to them. Rumi may inspire us to make that effort, he certainly inspired me.

I have chosen Gandhi as my third well of inspiration, for we live in a dangerous world in which enlightened leadership based on moral and spiritual ground is hard to find. We so badly need this kind of leadership to speak for the vast impoverished world; we need it to challenge the catastrophic arrogance, ignorance and lack of wisdom of our current leaders and the neglect of the deep and solid roots of our culture itself.

I think Gandhi was loved because he was sincere and because he behaved as an exemplar of his religious tradition, and felt viscerally the convictions of his people. Gandhi's unusual political and spiritual humility came about in part from his study of the *Bhagavad-Gita*, that profound and beautiful 'Song of the Master' that lies at the heart of the

Hindu tradition. He loved the *Gita* and carried it with him wherever he went. I want to take just one idea from the *Gita* that impressed and inspired me and was, I think, at the root of Ghandi's practical and genuine humility that served his people so well. In the third chapter of this epic poem, we find the verse that Gandhi particularly liked; 'Deluded is he who imagines he is the worker of his works.' In our world, which has so determinedly decided that we humans are the sum of all things, this sentence will barely pass credence. To Ghandi, it was the source of his humility.

Hindu sages of the past observed the deep powers of nature, the *gunas*, working through all creation, through all phenomena, including man. They saw the phenomenon of the creation constantly manifesting and falling back into itself, and they observed that man was intimately part of that process. You will find the same idea in the Chinese tradition of Lao Tsu, where: 'the ten thousand things arise by themselves.' And, of course, it exists in the Christian tradition, in the idea of the continual process of death and resurrection at the heart of daily life. So, are we harassed twentieth-century humans so sure we are the masters of our own over-managed house? I certainly arrogantly behaved as if I was.

Many great artists see their work as a gift and themselves as the instruments. Rumi says we 'are only the reed through which the Beloved plays his flute.' Mozart's music poured through him and he copied it down. Rilke's poetry swept through him like a tide and he copied it down. Some of the key details of Edison's inventions were simply revealed to him through dreams and visions and he copied them down. From whence did this inspiration come?

I finally became persuaded of this view in the events surrounding the attack on America in September 2001. The indirect effect of that attack was a deep fear that had perhaps been in the background all my life, which came to visit me with a vengeance. Through the gift of this great and cleansing fear, I came to see how not only was I really not the doer but also that I was part of something much more vast, part of a creation, part of a whole web of relationships that supported me; that in a sense held me. As the facts of my exterior life changed and as I came to see how there really was a security beyond the material in which I had deludedly put my sole trust, I came to see the limitations of the purely material.

Surely, this is the ground of a genuine humility that sees that we individuals are not the sole choosers and doers we think we are, for we live lives as part of that web in which the creator moves through us and the phenomena that arise in us. It is our job, not so much to choose what to do, but, in the words of the Lord Krishna in the *Gita*, to 'do the work that is laid upon you but to do it unto me, do it as a sacrifice to me,' or, to use a more familiar western text, 'to do the will of thy Father which is in heaven.'

The *Gita* has taught me much about my mind and about my origins. The Indians for centuries poured their genius into understanding the mind, while we in the West poured our equal genius into mastering the external phenomena of nature. Of course, the balanced person, or culture, needs both. But in our unbalanced state in the West, the *Bhagavad-Gita* and Gandhi's example have much to teach us.

Conclusion

Where does this leave me as I abandon these tossing seas I have journeyed through? Where does it leave me with myself and with those who come to me for advice and help? Jung taught me how to find a rich and precious interior life and filled me with a sense of meaning in the process of life itself that I had lost. Rumi reconnected me to the world of Beauty and the life-giving practice of working with internal imagery, that led to an intimation of who and what the Beloved might mean for myself and my world. Gandhi helped me to stop beating myself up against an impossible and cruel standard of material achievement and helped me to see how I really do not hold the heavens up all on my own, how there are unseen hands doing that for me, and showed me the gift of being a part of a creation that is in some manner sacred, whose manifold gifts include a security that is not only material. I can say that after much was beaten from me on the anvil of rough old life and after I began to gain a sense that I am part of a larger whole. As I continue to put one puzzled foot in front of the other on my journey, I notice that it is my foot that moves forward step by step and that what is developing with each step is a way, a path, unique and precious. My way, and yet, not of course really mine, for it is constantly being filled with the only thing that really matters — the immeasurable force of the profound mystery of life itself, a mystery that we all share.

Sky and Psyche, heaven and earth, visible and invisible, man and woman, always the gates of the opposites for us to travel through for our larger and more rounded and more 'whole-y' humanity, resisting, if we can, the attractive and comforting one-sided vision of certainty that is in the end. For, without these battles of opposite tensions, without connection to our root, to our own deep places, without the anchor of Soul, the world literalizes and loses its deep mystical root. Priests become welfare workers, rather than conductors of the Divine; poetry becomes a literal description of unconnected objects and personalized pain, rather than the language of heaven and the Soul; marriage becomes a utility, rather than an urgent quest to find, through your opposite, your own royal centre, your own Beloved.

And these battles can be won. This interior work slows up the pointless rush that is the curse of our age. It will painstakingly restore a space for beauty, a sense of who we are for each other, that we are part of a whole, that we have deep roots that reach beyond a house, or family, or a patch of dirt however much loved. This is the root that is referred to in Rumi's poetry as the 'root of the root' and is where our only security and that of our world can be sought.

I have no idea if any of these ideas have moved you to a place of inspiration. Certainly they have over the years moved and inspired me. They have also helped me see what it might mean to be more fully human and what it might mean to live more securely and perhaps more gently upon this beautiful Earth that is our home.

Notes

1 From the poetry of Mansur al Hallaj, in Lings, Martin, *Sufi Poems: a mediaeval anthology* (Islamic Texts Society, 2004), p. 34.
2 In Harvey, Andrew, *Light Upon Light: Inspirations from Rumi* (California: North Atlantic Books, 1996), p. 53.
3 *Ibid,* p. 17.
4 From the poetry of Mansur al Hallaj, in Lings, Martin, *Sufi Poems: a mediaeval anthology* (Islamic Texts Society, 2004), p. 28.

11. Understanding the Modern Disenchantment of the Cosmos

RICHARD TARNAS

It could be argued that the topic of this conference, the relationship of the heavens to the psyche, represents a cutting edge in both cosmology and psychology today — though of course most academics in these two fields would be stunned to know it. The modern mind has long assumed that there are few things more categorically distant from each other than 'cosmos' and 'psyche.' What could be more outer *than cosmos? What more* inner *than psyche? Are they not informed by fundamentally different kinds of principles, the one objective, the other subjective?*

But a multitude of developments in many fields, from post-Kuhnian philosophy of science to post-Jungian archetypal psychology, now obliges us to recognize that, of all categories, cosmos and psyche are perhaps the most consequentially intertwined, the most deeply mutually implicated. Our understanding of the universe affects every aspect of our interior life from our highest spiritual convictions to the most minuscule details of our daily experience. Conversely, the deep dispositions and character of our interior life fully permeate and configure our understanding of the entire cosmos. The relation of psyche and cosmos is a mysterious marriage, one that is still unfolding.

I am going to speak directly about 'Sky and Psyche,' following the title and focus of our conference. This is such a rich and important topic because the relationship between sky and psyche defines our world view. It defines the cosmology of a culture, and the cosmology of a culture is essentially the container of that culture. Everything that we do, say, and perceive takes place within a framework of assumptions and underlying principles that are ultimately embodied in the nature of the cosmos as our culture perceives it. In this sense, our cosmology is the most embracing vessel within which our human existence takes place. And when the sky and the psyche are seen as radically separate, this creates a very different world view for the human being to live within than a world view and cosmology in which the sky is ensouled.

Let me begin with a sentence that Jung wrote late in his life. It was in *Memories, Dreams, Reflections* in the chapter entitled 'Late Thoughts' — an evocative title for a man in his eighties who had lived his life as Jung had.

> Our psyche is set up in accord with the structure of the universe, and what happens in the macrocosm likewise happens in the infinitesimal and most subjective reaches of the psyche.[1]

Now, this view is something that Jung came to gradually. As we will see, for much of his life, Jung struggled with just how closely connected the structure of the cosmos and the structure of the psyche actually might be. And let me parallel that quote with a passage from Plato, from his last dialogue, *The Laws*, written at the equivalent moment in Plato's life. Here, Plato is reflecting back on the Sophist philosophers and the mechanistic physicists and cosmologists of the period that preceded him:

> The truth is just the opposite of the opinion which once prevailed among men, that the sun and stars are without soul. ... For in that short-sighted view, the entire moving contents of the heavens seemed to them only stones, earth, and other soulless bodies, though these furnish the sources of the world order.[2]

Here we have these two great figures in the history of Western thought; in a sense they are the two paradigmatic figures of the archetypal perspective. Plato stands at one end of the Western intellectual tradition and Jung at the other. Of course, Plato has Pythagoras and the pre-Socratics behind him, as well as the entire Homeric and mythological understanding of the cosmos, but he articulates the archetypal perspective in a philosophically potent, influential, and enduring way. This cosmology then goes through many metamorphoses in the course of history — including its profound eclipse in the modern era (though not in Romanticism), followed by its remarkable rebirth in depth psychology. Jung is at the other end of this arc, drawing, in different ways, on Freud, Nietzsche, Schopenhauer, and Kant, as well as the nineteenth-century mythologists. So, here these two figures stand, both at the end of their lives, making these important statements about the relationship of sky and psyche.

It is helpful to remember that Plato himself was reflecting on a period of disenchantment, the eclipse of an ensouled cosmology amongst Greek intellectuals, under the influence of the secular scepticism of his time and an increasingly mechanistic, materialistic world view. In a sense, in the one hundred years before Plato, the whole subsequent trajectory of Western thought was played out brilliantly in a condensed miniature form. So much of the drama of the modern and even much of the post-modern period was anticipated in that remarkable Greek awakening between the later sixth and early fourth centuries BCE.

Let us compare those two statements of an ensouled cosmology with a paradigmatic statement by a contemporary cosmologist of our own time, Steven Weinberg, whose book *The First Three Minutes* summarizes modern scientific cosmology. At the end of that book, Weinberg sets out what has become a famous summation of cosmic disenchantment: 'The more the universe seems comprehensible, the more it also seems pointless.'[3] Here Weinberg cuts directly to the heart of the modern scientific, mechanistic, disenchanted, dualistic perspective — a view that still reigns in our universities and in our scientific culture, though it has indeed been shifting, both subtly and, in certain ways, quite dramatically.

In the scientific and philosophical revolutions of the sixteenth and seventeenth centuries, as well as the Enlightenment and the Darwinian revolution of the eighteenth and nineteenth centuries, the basic structure of the modern world view was brought forth. In this great shift, an unprecedented radical dichotomy was established between the human self and its interior world as subject, and the outer world — nature, the cosmos — as object. This dichotomy is unlike anything found in any world view other than the modern. All pre-modern, non-Western, and traditional indigenous world views in some sense recognize that meaning and purpose pervade the universe. Conscious intelligence, soul, and spirit are seen as not limited to the human being but as something larger that is participated in by the human being. But with this profound revolution of the modern mind, all meaning and purpose in the cosmos is relocated inside the human self, while the rest of the universe is voided of any capacity for the expression of meaning and purpose.

Let us consider the difference between the primal world view of *participation mystique* and the modern world view based on the subject/object dichotomy. When a person from a primal or shamanic culture walks out into the forest, he or she experiences spirits in the forest; there are psychic and spiritual presences in all forms of nature, in the wind, the water, and the sky. For a Native American or an African Bushman, for our own ancestors in ancient Europe, when two eagles are seen flying across the horizon or when two planets come into conjunction or when the Sun and the Moon move into alignment, these events are regarded as having significance. They bear meaning, and the reason that they can do so is that the cosmos and nature are recognized as being in some sense pervaded with meaning and purpose. They are capable of carrying and embodying meanings and purposes that are continuous with the meanings and purposes that the human being discovers inside herself or himself.

By contrast, in the modern world view, if you see meaning or purposeful significance in two planets coming into conjunction, or in two eagles crossing the horizon, this is the sign of a basic epistemological error. You are projecting human categories onto the non-human world and this is a sign of mental weakness, the act of an immature mind. It reflects that a kind of childishness has not been outgrown. In fact, if one persists in this kind of inappropriate projection, this could be a sign that the services of a psychiatrist should be sought, and that the person so afflicted should probably be put on medication. So this difference in world views is not just theoretical; it has concrete consequences.

There is an exemplary story in a small book of vignettes and memories about Jung, his wife Emma, and Toni Wolff that I happened upon at the Jung Institute on the shores of Lake Zurich. Some time in the 1950s, a man named Henry Fierz had come to see Jung to discuss the possibility of publishing a book that had been written by a scientist who had recently died. They met at five o'clock at Jung's house. Fierz describes the scene:

> Jung had read the book and he thought that it should not be published, but I disagreed and was for publication. Our discussion finally got rather sharp, and Jung looked at his wristwatch, obviously thinking that he had spent enough time on the matter and that he could send me home. Looking at his watch he said: 'When did you come?' I: 'At five, as agreed.' Jung: 'But that's queer. My watch came back from the watchmaker this morning after a complete revision, and now I have 5:05. But you must have been here much longer. What time do you have?' I: 'It's 5:35.' Whereon Jung said: 'So you have the right time, and I the wrong one. Let us discuss the thing again.' This time I could convince Jung that the book should be published.[4]

This story illustrates two important points. One is that Jung was constantly alert, like a shaman, to the signs that nature or the immediate environment, the non-human environment, is always giving to the human being. The unexpected stopping and error of the watch was immediately recognized by Jung as paralleling — and as thereby bringing to his attention — what he then suspected might be a comparable stoppage and error in his own thinking about the issue at hand. He saw the two events as coherent, as reflecting a larger pattern or field of coherent meaning. While most people would simply think, 'Damn, my watch is broken again!' Jung saw a meaningful pattern and responded to that perception.

We know from the reports of others that Jung lived this way all the time. When he was in session with a patient, meeting in his house near the lakeside, he would notice when the lapping of the waves got louder, or when the wind became stronger, or when a flock of birds suddenly landed on the lawn. He repeatedly observed how such events closely corresponded to the events that were happening inside the consulting room, in the interior life of the patient and in their dialogue. Of course, there is the famous example of the golden scarab flying in the window at the crucial moment when the woman is describing her dream about a golden scarab, that classic synchronicity that he gives in his monograph 'Synchronicity: An Acausal Connecting Principle.' Now this kind of alertness to the signs that synchronicity can give us would be viewed by the modern mind as simply pathological. A modern psychiatrist would, as a number of psychoanalysts actually did, diagnose Jung as being an ambulatory schizophrenic.

The other crucial point in evidence in the story of the stopped watch, beyond the fact that he observed the pattern and found it meaningful, is how Jung interpreted that sign for his own self-correction, for compensating his one-sided conscious attitude. He didn't use the coincidence that was focused on his actions to build up his own sense of self-importance, like 'the whole world is centred on me, everything is meaningful and purposeful according to what I need and want.' Instead he used the coincidental pattern to correct his own ego's one-sided assertion of its will, because he saw a larger context of meaning containing a striking parallel between the stopping of the watch and the stopping of his own mind happening at the same time. He used the hints from the world to correct the one-sided blindness of his conscious attitude. As Jung emphasized, a properly interpreted synchronicity will usually be a defeat for the ego in favour of the deeper Self. This is very different from how some New Agers tend to use synchronicities, to narcissistically build up their sense of self-importance and self-centeredness.

Jung's perspective about synchronicities and his shaman-like alertness to such meaningful patternings is remarkable. Jung's published writings do not reveal the actual extent to which he was using synchronicities in his daily approach to life. I think the reason for this withholding on his part, much like his many hedgings and ambiguous statements about astrology, is that he knew he had pushed the envelope about as far as you could push it in the mid-twentieth century. Yet, more than this, I believe that Jung was genuinely struggling with the power of the modern disenchanted universe that he himself had deeply, if not completely, assimilated.

Let us recall the quotation that Jules Cashford read from Jung, which is so representative of many statements he made throughout his career up until the late 1940s:

> All the mythological processes of nature, such as summer and winter, the phases of the Moon, the rainy seasons, and so forth are in no sense allegories of these objective occurrences; rather they are symbolic expressions of the inner, unconscious drama of the psyche which becomes accessible to man's consciousness by way of projection — that is, mirrored in the events of nature.[5]

Jung would often emphasize that it is a sign of how deeply-rooted this perceived connection or conflation is between the outer forms of nature and these inner symbolic processes that it took so many centuries to differentiate and disengage the self and the projecting mind from those processes. That is why he called astrology the summation of ancient and medieval psychology, stating that it represented a projection of the mythic forms of the collective unconscious onto the heavens. He made many statements along that line, outside of astrology as well. These are all signs of the degree to which Jung was struggling with living in the modern disenchanted universe while absorbing the meaning of his many epiphanies that suggested matters were rather more complicated. It was really not until the last ten or fifteen years of his life that he became more and more clear that these archetypal principles and forms are not just intrapsychic, in the sense that they are rooted ultimately within the human psyche. And he starts using the somewhat awkward word 'psychoid' — to convey that the archetypes are psyche-like, that they are both inside and outside, that they inform both spirit and matter. The concept of synchronicity for Jung thus represents a way in which the modern disenchanted universe starts to be broken through and dissolved. I believe that's why Jung expended such extraordinary effort to bring the concept of synchronicity into modern discourse, because he recognized that this is a crucial way of breaking out of those 'mind-forged manacles' (to use one of Blake's evocative phrases for the modern disenchanted vision).

The negative consequences of having a disenchanted universe and a separation of sky and psyche have been recounted many times. So I will just quickly sum up what I would view as being the major negative consequences. But then I am going to go past that because I think we need to recognize that history is not a simple melodrama of good versus bad. History has a profoundly tragic dimension, but it is not a melodrama. It is too full of paradox and complexity and ambiguities for there to be just a good side and a bad side, or to believe that everything was great until one terrible thing happened and then the rest is a great fall. Something more profound is going on with the emergence of the disenchanted perspective, of the modern mind and the modern self. But first let us quickly summarize the problematic side of disenchantment.

First, of course, we have the alienation of the modern self in a cosmos that has no coherence with our inner spiritual aspirations, with our psychological nature. We are not at home in such a cosmos. We are a stranger in a strange land. There is a kind of 'darkening of the world,' to use Heidegger's phrase. If we live in a world of random purposelessness and meaninglessness, this casts into question our own purpose and meaning, our very essence.

Moreover, a second and related negative consequence of the disenchanted perspective is that there is a loss of metaphysical context, a loss of the cosmic meta-narrative that provides a coherent framework of intelligibility for human life; these are cascading and overlapping problematic outcomes of the disenchanted world view. We also see that disenchantment brings a loss of a cosmic foundation and an encompassing context that would provide a basis for spiritual, moral, and aesthetic values. This foundation disappears if you live in a meaningless, purposeless, random universe of matter and energy alone. If the cosmos and the nature of things does not provide a larger basis of motivation and values beyond the narrow utilitarian impulses and values of financial profit, political power, and technological prowess, then those utilitarian values are going to rule the society — and that is exactly what is happening today.

In such a universe, the natural world simply does not deserve moral concern. Trees just become lumber. Animals are harvestable commodities. Children's minds are marketing targets. The most beautiful work of music by George Gershwin can be shamelessly appropriated for twenty years as an advertizing enhancement for United Airlines. Everything can be appropriated for profit and power, because, quite literally, in a disenchanted universe, nothing is sacred. And, of course, we can see that, from this cascade of negative consequences, a disenchanted objective world is ultimately going to rebound on the human subject, by disenchanting the human self. If we are evolved from a meaningless world, then our claims for deeper meanings and purposes are undermined right out from under us. We become just a statistic, a commodity, a genetic strategy for the selfish gene, a meme machine, a mere sociobiological program.

And then there is the larger loss of the sacred, for when you lose it in the cosmos and nature, then you begin to lose it inside as well. It becomes harder and harder to support the interior if the exterior has been lost. And finally, in a disenchanted world you also lose the context within which the great initiatory rites of passage always took place. Initiations always served in a society to heal that rupture of Being that takes place in human existence, but such a ritual is meaningless in a meaning-empty cosmology. So a feedback loop happens: The cosmology undermines the *raison d'être* for the initiatory rite of passage, and then you have generation after generation of uninitiated individuals — scientists, philosophers, religious authorities, everyday people — who do not perceive an enchanted universe because they have lost it and have not been initiated back into a more profound access to it.

We could say that these are all forms of what Max Weber called the 'iron cage' of modernity. And of course the word 'disenchantment' as I am using it comes from Weber as well, who was developing an insight of Schiller's a century earlier. But what about the positive side of disenchantment? Why did disenchantment happen at all? What factors were involved in this tremendous transformation?

I believe that we first need to recognize that this was not just something we can lay at the feet of, say, Descartes or Bacon or Locke, and single out that individual as being the source of the whole problem, the villain of the whole melodrama. Actually, in certain ways Descartes and Bacon and Locke — like Darwin and Nietzsche, Marx and Freud — each one of these disenchanters was in certain respects a profoundly liberating figure. They were all very courageous protagonists in this drama, and in many ways we are deeply indebted to them. But, above all, the point I want to make here is that these figures all reflect larger processes at work. Somebody like Descartes is not just coming up with a statement like *cogito ergo sum* out of the blue and then everybody thinks in those terms from then on. By making that sharp distinction between a rational self and an objective world that is different in character from the perceiving subject, that we can manipulate and exploit through our rational understanding, Descartes is articulating an emerging sensibility that the entire collective psyche of the modern world was bringing forth. He is somehow the spokesperson and vessel for this larger transformation in the same way that, several hundred years later, Jung was the vessel for something new of a very different sort. So we need to move past these reductive and simplistic personal understandings to a more transpersonal understanding.

As for the positive factors involved in modernity's cosmic disenchantment, first, there was a tremendous overcoming of inherited tradition-bound, pregiven structures of meaning and purpose that were allegedly built into the very nature of things and into the nature of the cosmos. These pregiven structures had often been interpreted and enforced by religious and political authorities whose profundity of insight might at times have been justifiably questioned and whose motives might not have always been pure. Often, the enchanted world view was spun through local nationalist, tribal, and religious mythologies that wrought tremendous destruction, as is happening in our own historical moment. So disenchantment plays a crucial role in liberating the modern mind from that kind of bedazzlement and possession.

Second, there is a tremendous increase in individual autonomy that takes place with disenchantment. Even though a culture's established world view might provide a sustaining quality of luminosity and meaning to its people, when there is no change and no possible change for generation after generation in that world view, the constraints on the individual's freedom of thought and action are manifest. How precious it is to us living today that we can question our world view, and that we can question that what our parents, small town, country, or religion taught us is true! This freedom is, generally speaking, a modern phenomenon. In our own age we can go through several world views in a single life time. In California it can be several world views in a single weekend. And this autonomy is something that we deeply value. In fact this may have been part of a *telos*, a deeper purpose, in disenchantment: It both permits and obliges the individual to enter into an authentic, original relationship to life's fundamental conditions, one that has not simply been force-fed to everyone in the culture. In some sense you are now potentially able to play a key role in the larger scheme of things, in that you have to use your own intelligence and courage and your own imagination, and follow your own passions and pursue your own quest, in order to freely play a role in how the universe unpredictably unfolds.

Returning now to the seventeenth and eighteenth centuries, another related factor in why disenchantment took place is that when the external world is objectified, the human subject is in certain respects empowered. Making the world an object creates greater subjective potency. This is a very important point that was clearly articulated by the Canadian philosopher Charles Taylor, particularly in his brilliant book *Sources of the Self: The Making of the Modern Identity*, but also in his earlier excellent book on Hegel. In the latter, he makes the point that prior to any technological 'pay-off' of the mechanistic perspective and disenchanted world view, this austere vision brought a kind of exultant euphoria to the modern self, because there was now a new capacity for self-definition. I could now define myself not according to a pre-given set of meanings that were embodied in the universe, with which I needed to align myself in order to realize myself. Instead, that world was in some sense neutral, and I could then define myself with greater latitude of freedom — a new autonomy which all of us now draw upon and assume in our lives.

Yet another major factor in why the modern mind felt so confident about disenchantment is that it came out of that extraordinary cosmological event called the Copernican revolution. The Copernican revolution gave the modern mind a confidence that was beyond anything it had ever experienced. This was the very first time since the ancients that there was a widespread feeling in the intellectual community that it understood the actual nature of the heavens. Up until that time, the entire period from Eudoxus and Ptolemy right up to Copernicus and Kepler, the most that could be sought were instrumentalist fictions, mathematical constructs that would give an approximate rendering of the planetary positions. But because astronomers were working within the constraints of the geocentric cosmology, the predictions of planetary positions based on these constructs were constantly inaccurate. Astronomers therefore did not believe that their equations and calculations were actually showing the true planetary motions, quite understandably when you look at all those epicycles and eccentrics and equants that had to be assumed in order to produce the predictions. But finally, with Copernicus and Kepler, astronomers could see how the orbits of the planets and the Earth moving in ellipses around the Sun actually could be physically moving, continuously and regularly, in the way their mathematical calculations suggested. So, there was now an extraordinary sense that the modern mind had actually grasped the cosmic order of the divine mind, that an unprecedented illumination had happened. Copernicus and Kepler had truly fulfilled Plato's call to solve the 'problem of the planets' — their apparent retrograde motion and chaotic wanderings — and in doing so had at last comprehended the divine mathematics of the macrocosm. When one considers that up until that point virtually everyone since the birth of humankind was firmly convinced that the Earth was stable, fixed, central, self-evidently unmoving in its great vastness — and then suddenly an entire civilization wakes up to the fact, the conviction, that the entire Earth is actually moving through the heavens, in many ways at once: such a cosmic shift is virtually the most fundamental fact one can imagine. Suddenly every other culture, every other era, every other mode of knowledge, every religion including Christianity itself, was now seen as being in some fundamental sense inferior to the unprecedentedly powerful rational intelligence of the modern self.

Here we can see how the rational mind of 'modern man' identified itself with the Sun, with the Solar Logos of the divine reason. The modern self, therefore, did not feel that it was being absolutely de-centred by the Copernican revolution (as a simple account would have it), because it saw that the modern mind, like the ancient Solar Logos and the divine mind, shone brilliantly on and illuminated the entire universe through its reason. It illuminated the universe in the same way that the Sun did, and so the modern mind subliminally felt that it was a kind of embodiment of the Solar Logos that had never been seen before. And a certain arrogance and hubris crept in because such a sense of intellectual superiority over all these other forms of knowledge and other eras had been so vividly demonstrated on a cosmic scale for all time.

Related to this euphoric solar elevation is the sense of existential liberation that the modern self soon experienced with the post-Copernican universe's negation of the ancient and medieval reality of the Christian hell. It is perhaps a little difficult for us today to recognize to what extent in the Western soul the fear of hell, as a potential fate for each of us and for the majority of humankind, loomed over human existence, day after day, year after year. For hell was envisioned as located at the dark centre of the fixed earth, just as the divine was seen as surrounding the outermost luminous circumference of the cosmos. The further away from the Earth, the more celestial, the more divine, the closer to God. The closer to Earth, the less divine, the more mortal and imperfect, the more fallen. And at the very centre of the Earth, in its 'bowels,' was hell and the power of Satan. Dante's *Divine Comedy* is of course the great poetic summation of this cosmology.

According to the dominant perspective of the biblical tradition that had developed from the time of the ancients and Augustine through the medieval period, most people were destined for hell — indeed predestined for hell. But as the Copernican revolution unfolded, it was discovered that while there did not appear to be any angels in the heavens visible through the telescope, nevertheless if the Earth was moving, and was itself in the heavens, then hell with its teeming devils and eternal damnation was also called into question. Hell was suddenly removed from its ancient position deep underneath us at the Earth's centre. An enormous sense of existential emancipation emerged with this realization. Original sin — that sense of being so unspeakably unworthy and so close to eternal damnation at every moment because all you would have to do was think an erotic thought for the dark sin to be recorded and the horrible destiny mandated. If you think I am exaggerating, reread the brilliant third chapter of James Joyce's *Portrait of an Artist as a Young Man*, with its spectacularly eloquent, richly-imagined sermon by the Irish Catholic Jesuit retreat master about the nature of hell. Joyce could not have written that sermon unless in a fundamental way he had been able, like the modern self, to free himself from the power of its dark vision, because it sums up superbly every sermon on hell that has been given from the Middle Ages to that moment. So the Copernican revolution, which broke through the ancient division between the celestial and the terrestrial and the ancient devaluation of the 'sublunary' world of humankind, overthrew the most extreme form of that dichotomy with the disappearance of hell at the centre of the Earth.

So that is the religious liberation that emerged out of the scientific. Certainly much of the passion which motivates defenders of a strict Darwinian perspective derives straight from that same lineage, because if Darwin is right then there isn't this eternally-damning God leaning over our shoulders ready to send us to hell. It's worth noting how frequently the most fundamentalist scientists, particularly the fundamentalist Darwinians, seem to have undergone a strict, fundamentalist religious upbringing. They are driven to defend the new dispensation in terms equally dogmatic to the religion they despise, for they are rebelling against their parents, ministers, and priests. They feel that their most essential freedom rests on this disenchanted scientific world view being absolutely correct, and they haven't envisioned any other possibilities.

They are still holding on rigidly to an earlier moment of transformation, like a traumatized adolescent, and not seeing that the game has become much more complex.

If we enter more deeply into our history and the evolution of our world view, we can begin to appreciate the extraordinary paradox and rich complexity that pervade it. And one of the ways we can go more deeply into it is to recognize both the many problematic and many positive dimensions of that evolution. Of course, we have not mentioned all the empowerments, conveniences, and securities that have come with this modern development. Consider the sense of increased safety that modern science has provided us with in relation to so many diseases that have afflicted humanity, or consider the sharply increased safety of childbirth. All one has to do is walk into any cemetery and see how early in life many women died, and how many babies died at birth, to know that certain advantages have come from the mechanistic medical perspective — not that this is the last stage of development, but extraordinarily positive things have indeed come from it.

With all these positive and problematic consequences of disenchantment, I believe we are seeing a long, gradual evolution and forging of the human self, with the modern self a kind of evolutionary experiment in which we are all participating. Those of us who seek to have an 'integral' world view, if we are to be consistent, cannot say that the whole universe is one, and the world is ensouled, and everything has a unitive dimension and participates in the whole *except* for the modern mind, or reductionistic modern science, or patriarchy, or the disenchanted world view — everything is part of the one, except somehow this particular thing is not. For this too must be part of the one.

I believe we can acquire a deeper insight into the psychology of disenchantment if we keep two points in mind: First, it didn't just happen out of the blue in the sixteenth and seventeenth centuries. It really rests on an enormous evolutionary shift that began probably as early as the emergence of *Homo sapiens*. As soon as you start having tool use, as soon as you start having linguistic symbolization, you start having a subject/object dichotomy. But it is in the great transformation of religious perspective, that happened in the first millennium BCE right across the world from Asia to the Mediterranean, that we see a series of stages that move human consciousness ineluctably towards disenchantment. This brought a new sense of the divine and the ultimate locus of meaning and purpose as transcending this world — which in the West culminated in Christianity. Early Christianity and Roman Catholicism to a certain extent still possess a great deal of enchantment, as did biblical Judaism. The world still is a theophany, but gradually the perspective that 'man was made in the image of God' in a way that the rest of the universe was not, and the understanding of human history as a locus of divine activity in the world that is far more important than the eternal cycles of nature, these shifts work more and more on the collective psyche in the West until the eruption of the Protestant Reformation takes place at the same time as the beginnings of the Copernican revolution. In the Reformation a militant desacralization of the world takes place, whether it is against magical ritual in the churches or against lingering nature spirits or against the Virgin Mary and saints and relics; all the enchantments and remnants of the ancient sacralized world view lingering in medieval Christianity are now vigorously suppressed. Calvin is of course the key figure in this change. Why the militant vigour of this desacralization of the world? Because then the glory of God, this transcendent separate Being above the world upon whom man is modelled, could be better served by man, and the world could be transformed to better serve the divine glory. And that desacralization provides a crucial step towards the modern scientific world view.

The Protestant Reformation is key to the larger sequence, and we can get a glimpse into this by a very interesting statement that the influential evangelical preacher, Billy Graham, once made. Sometimes you hear a person say a single sentence that suddenly illuminates their larger vision like lightning in a dark landscape. I happened to see this on the television news in the early 1990s, when Graham was visiting a place in the Carolinas where many people had just lost their homes and their communities to a powerful hurricane. In the process of consoling the bereft people, he said: 'If you don't have faith in God the Father, you are entirely at the mercy of Mother Nature.' In that wonderfully revealing statement you can see the whole collective psychology of the evolution of consciousness as it first moved away from the reliance on, and immersion in, an ensouled nature of spirits and demons and deities, instead resting one's faith in a supreme transcendent God who is beyond this world where one's soul might someday transcend the sufferings of this life, which is a mere prelude to the afterlife. We can see a new secularized form of this statement in: 'If you don't have faith in modern science — Father Science, as it were — then you are entirely at the mercy of everything from the Mother Church (and religious superstition and ignorance) to Mother Nature (and its destructive forces and diseases).'

In the gendering of those statements, in the metaphors that are used, we can see how much these dichotomies and transformations are related to the epochal shift from a more gender-neutral or feminine-informed perspective, where one has animism, polytheistic deities, or a powerful all-pervading Goddess, towards a series of distinctly masculine symbolizations of both the divine and the human: Yahweh, God the Father, Zeus, Prometheus, Christ, Man, 'modern man.' And this can be seen as a vast evolutionary movement from a more lunar perspective, lunar and earth (sublunary) together, to one dominated by the transcendent solar hero, the Solar Logos, the Sun. The Copernican revolution was the cosmological ratification of this vast solar development, bringing the Enlightenment and the modern self in its wake.

Many of the contributors to this conference have discussed the relationship between the lunar and the solar. How are we to balance our knowledge of the Sun's greatness, its centrality, its magnitude, its life-giving power, its illuminating power, with our awareness of the Moon, which particularly in our patriarchal culture and our modern scientific culture is obviously an inferior entity, much smaller, without its own energy source or light, merely reflecting the Sun, merely revolving around the Earth? It would seem that this is such an imbalance. How can we have a genuine solar/lunar balance? Likewise, how can we have a masculine/feminine balance? I believe that the way to move beyond that imbalance is to recognize that the Moon is not simply that small satellite of the Earth. It is no accident that in our perception from the Earth the Moon is virtually identical in size to the Sun, so that when they are perfectly aligned in an eclipse a magnificent cosmic event is created that leaves scientists, as well as everyone else, in awe.

Let us see the deeper archetypal parity of the Sun and Moon by recognizing that the Sun is the ruler of the day and the Moon is the ruler of the night. When the Sun shines, it illuminates all with a light that completely overpowers the visibility of the rest of the sky. By contrast, when the Moon shines one can still see the rest of the sky — all the other stars, all the other lights, the entire *anima mundi* of the night sky is visible. The Moon is the ruler of the whole, and the night is pregnant with the day. The night gives birth to the day. The lunar gives birth to the solar. The night sky is the whole dark fertile void of the cosmos out of which all stars and all suns emerge, and every day comes out of that fertile, transformative night. The Sun is the part of the whole that in a sense heroically carries forth in its specificity the whole *in potentia*, while the Moon is the whole that like a mother contains the part within itself *in potentia*.

Our whole image of the hero as being basically a conqueror, one that is trying to assert itself over the whole, which has come in for such deserved criticism in recent decades, is really an image of the immature solar hero, growing and moving up in its ascent to the Midheaven. But eventually the Sun descends. The Sun has to go down, just as the true hero must go down, and in many ways we are at that extraordinary moment in our civilization's history when the Sun is going down. Nietzsche is the one who recognized this first and most powerfully. He called upon those who would be true to the hero in their souls to have the courage to go down. The Sun must go down for there to be new dawns and new horizons. For the full trajectory and arc of the human journey to unfold, the solar self has to descend into the lunar ground to be transformed, and this takes a tremendous act of courage.

I believe that astrologers are in a unique position at this historical moment. They occupy a key role in this transformative moment in the heroic solar journey of the West and of modernity. In certain ways the myth of Western civilization is precisely this Promethean solar trajectory with all the shadow that goes with the solar light, and the need to acknowledge this shadow, suffer its consequences, and integrate it. It is timely now to say it, that one of the most important reasons why the modern mind's disenchantment of the universe was so liberating was that astrology had become so confining. The power of the planets to fatefully determine human life was already emerging in the later Hellenistic and imperial Roman period, which was becoming more and more deterministic and mechanistic in its understanding of the cosmic powers. Many factors went into this but one can see this general tendency on many fronts at that time — in Stoicism, Gnosticism, and the mystery religions. This is why so many of the latter taught that the planetary powers were the ones you needed to go beyond, to transcend. They were confining forces, and the older understanding of the planetary powers as powers with which one had a dialogue had been lost. Astrologers were predicting what was going to happen to you fatefully, and are still doing this. This is precisely why Christianity had such an ambivalent relationship with astrology, and why modernity entirely rejected it. Both were championing human freedom. This is part of astrology's shadow which must be faced and assimilated if astrology is to emerge again into the centre of culture where it belongs.

We have reached an extraordinary moment in our history. The emergence of depth psychology and the tremendous intellectual-psychological-imaginative development that has taken place in the last century, and the larger development of the last several centuries that has resulted in the forging of the autonomous modern self, all this allows us not only to recover the *anima mundi* but to re-engage it in a new way, whereby we can co-creatively participate in its unfolding rather than simply deny it, try to control it, or be oppressed by it. We have all the intellectual and imaginative preparation now, such as the understanding of symbol and archetype over the literal and univocal, the understanding of archetypal multivalence. Archetypes can express themselves in many different ways and still reflect their essence, and we can play a role in affecting and inflecting how the archetypal powers will unfold in our lives. From this perspective, the human spirit and the human imagination is the cosmos's spirit and the cosmos's imagination working in us and through us. We are now at that key moment where we can again come into *relationship* with the whole, with the *anima mundi*, with the universe, as an 'I to a Thou' (to use Buber's great formulation), rather than an 'I to an It.' And in that newly co-creative and participatory relational engagement of a courageously autonomous self with the cosmos, and with the *anima mundi,* we have the possibility of a true Solar/Lunar *conjunctio* and a true sacred marriage.

1 Jung, C. G., *Memories, Dreams, Reflections* (New York: Pantheon, 1963; Vintage, 1989), p. 335.
2 Plato, *The Laws*, A. E. Taylor, trans. (Princeton, N. J.: Princeton University Press, 1965), XII, 967c, p. 1512.
3 Weinberg, S., *The First Three Minutes: A Modern View of the Origin of the Universe* (1977), 2nd ed. (New York: Basic Books, 1993), p. 154.
4 Fierz, H., in Ferne Jensen (ed.), *C. G. Jung, Emma Jung and Toni Wolff: A Collection of Remembrances* (San Francisco: The Analytical Psychology Club of San Francisco, 1982), p. 21.
5 Jung, *Collected Works*, vol. 9, p. 6, in Jules Cashford, *The Moon: Myth and Image* (London: Cassell, 2003), p. 142.

12. The Secret Life of Statues

ANGELA VOSS

The ancient ritual art of telestike, *or statue-animation, depended on a mode of perception in which divinity and matter were united through the action of* psyche, *or soul. In this paper I will explore the power of statues as magical baits, to attract the life of the cosmos and awaken the active imagination of the human soul through its erotic desire for union with beauty. As an image of the divine feminine or the queen of heaven, Psyche lives in the sky, exiled from the realm of human nature. I suspect that she is kept there, deprived of her sensuality, by our loss of ability to trust our innate perception of material images as alive, and also by our allegiance to both the ambivalence of orthodox religion towards human sexuality and the modern rationalist distrust of the imagination as a valid mode of knowing reality. My aim is to evoke an alternative kind of vision, a re-claiming of the validity of instinctual and intuitive response, and to do this, I shall draw on a variety of cultural, historical, philosophical and poetic perspectives on the function and purpose of representational art.*

To begin, a poem by the German lyric poet, Rainer Maria Rilke, 'Archaic Torso of Apollo':

> We cannot know his legendary head
> with eyes like ripening fruit. And yet his torso
> is still suffused with brilliance from inside,
> like a lamp, in which his gaze, now turned to low,
>
> gleams in all its power. Otherwise
> the curved breast could not dazzle you so, nor could
> a smile run through the placid hips and thighs
> to that dark centre where procreation flared.
>
> Otherwise this stone would seem defaced
> beneath the translucent cascade of the shoulders
> and would not glisten like a wild beast's fur:

> would not, from all borders of itself,
> burst like a star: for here there is no place
> that does not see you. You must change your life.[1]

As Rilke contemplates the statue, he becomes aware of a life, or a power, within it — an inner brilliance which sees *him*. This experience of recognition, of direct contact, makes him suddenly aware that he must change his life. Why should this be? How can a work of art come alive like this and affect him so deeply? His reaction raises the question of what it means to see images and statues in a different way from a conventional aesthetic appreciation; a way which involves staying with both the recognition of their inner life and our emotional (that is, imaginative and erotic) response to it. The metaphor of awakening Psyche from her slumber points to the emergence of a clearer, more focused mode of perception as we allow the intention of our imagination to animate the image. What is more, the waking is reciprocal, as we allow Psyche to open *our* eyes and lead us to a place where we can unite with her. In other words, I am proposing that the force of attraction and desire can infuse soul into matter and effect a union of the two worlds we call spiritual and material. But we have to dare to allow this to happen, for our response as post-enlightenment rationalists is inevitably conditioned by the reflex of 'common sense' which provides a sturdy defence against the fear of losing intellectual control over our objects of scrutiny. Profound experiences of love and desire are especially prone to denial, a reaction explored by the Jungian psychologist, Aldo Carotenuto:

> Whenever one rejects the experience of love by rationalising it away, one is obeying a collective law that has been internalised. We have all absorbed this law that negates the free realisation of desire in the face of life's continuous invitations. Thus while life conspires to arouse us, it can — and does — often happen that we deny our desire in obedience to an external veto that by now is fatally alive within us without our even being conscious of it.[2]

Furthermore David Freedberg has pointed out in *The Power of Images*: 'we go into a picture gallery, and we have been so schooled in a particular form of aesthetic criticism that we suppress acknowledgement of the basic elements of *cognition* and *appetite*, or admit them only with difficulty.'[3] However, he suggests that sometimes we are confronted with such a powerful fusion of an image and its prototype (and this is particularly the case with statues) that we cannot mentally detach ourselves and sense a living presence which disturbs, attracts and threatens to shake the literal foundations of our reality (see, for example, Plate 1).[4]

Not all representational art will be capable of stirring us in this way, and we are inevitably subject to constant anaesthetization by the creations of the media, whose soul-less, fabricated glamour vanishes as soon as it appears. But as the writer Ursula Leguin notices, the mysterious power of real myths cannot be destroyed by our reason (or our cynicism) however hard we may try, whereas fake ones are seen straight through. 'You look at the Blond Hero' she says, 'really look, and he turns into a gerbil ... But you look at Apollo, and he looks back at you ... [Figure 1] When the true myth rises into consciousness, that is always its message. You must change your life.'[5]

How, then, may we allow the 'true myth' to reveal itself? There are all sorts of ways in which the rational mind stops us at the brink of

*Figure 1: Head of Apollo, from the west pediment of the Temple of Zeus at Olympia (*The Greek Museums, *Ektodike Athenon, 1975), © Archaeological Receipts Fund*

engaging with Psyche. We may retreat to 'image theory' or transfer our attention to historical and cultural contexts or the biography of the artist. Psyche is then kept at a distance, or worse, suffers the indignity of being reduced to an irrational aberration — for when our intuition of living presence in an image is dismissed in favour of the reassuring knowledge that it is 'only' stone or paint, it easily becomes relegated to the domain of superstition or groundless fantasy. If, however, we *trust* our instinctual response, the material dimension becomes part of the paradox inherent in any symbolic image; it is made of inert substance, and *yet* it points to something alive. After all, if statues or paintings of provocative subjects were merely inert matter, why would they be censored? Why would erotic imagery be considered dangerous or corrupting? It arouses us because the outer form is 'seen through,' to something deep, immensely powerful and possibly threatening to civilized human behaviour, something which stirs our appetite for possession and dominance under the guise of love.[6] As we have no place, no direction or spiritual channel in our received religious tradition to work with this highly charged, often overwhelming, emotional energy, we are confused, ashamed of the exposure of that raw place. Added to which, our Christian legacy of the last two thousand years has left us wary about the nature of sensuality and desire, and instead of allowing Eros to reveal his divinity through the arousal of our senses, we separate soul from body, sacred love from profane passion, divinity from matter. In *The Erotic Word,* David Carr discusses the separation of sexuality and spirituality which runs through the heart of Western culture — particularly Western Christian culture — leading to an assumption that spiritual experience is somehow beyond, or superior to, the longings of the flesh. He has championed a reading of the Bible which revisions human sexuality as an image of divine creative power, and erotic passion as the animating force of creation which leads human beings to connect with each other and with God.[7]

Certainly figural art in general — and figural sculpture in particular — has been the most powerful medium for human beings to re-create themselves in the image of their deities (or to create gods in their own image) as a way, perhaps, of revealing their innate divinity to themselves as in a mirror. Just as the God of Genesis formed man from the dust of the earth and breathed divine life into him, we as humans have, from time immemorial, created beings — in stone, clay, wood or metal — which surely only await the breath of our passion to soften their carved or chiselled marble flesh (see Figures 2 and 3).

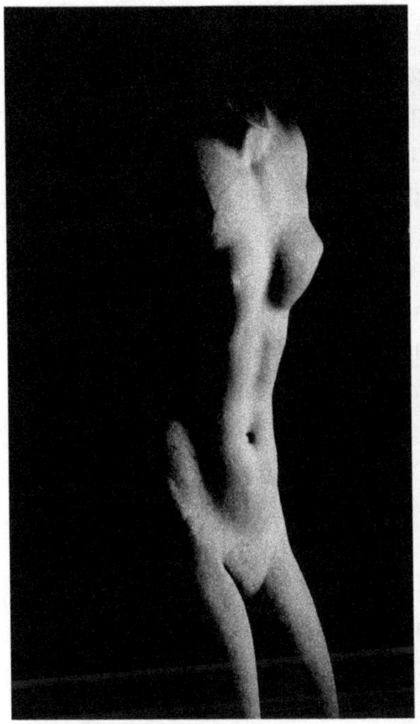

Figure 2: Sinuessa Aphrodite (marble), Museo Archaeologico Nazionale, Naples, Italy, (Bridgeman Art Library)

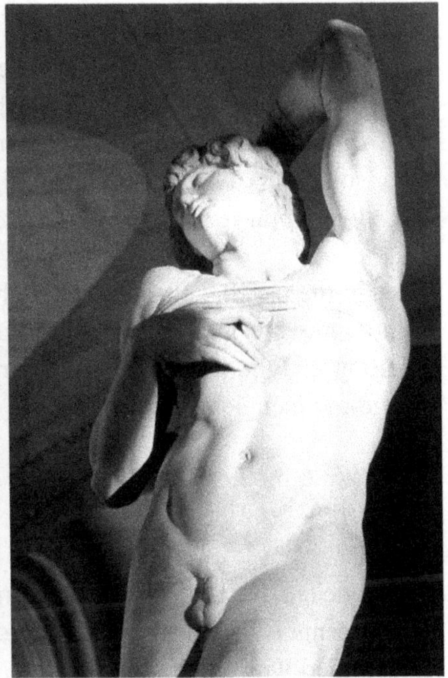

Figure 3: Michelangelo Buonarrotti, The Dying Slave, *photograph copyright Robert Moody*

The ancient Greek and Renaissance sculptors knew how to cast spells in marble, inviting us to suspend our disbelief in its warmth and pliability. How could we not desire to touch and caress these bodies? Now what would happen if we allowed our attention to focus, our desire to grow, our longing for contact to reach beyond the sensual and intensify? We move into the territory identified by C. G. Jung as the active imagination, the technique of concentrating on an image (internal or external) so vividly that it begins to live, or become 'pregnant' with possibility:

> *Looking*, psychologically, brings about the activation of the object; it is as if something were emanating from one's spiritual eye that evokes or activates the object of one's vision.
>
> The English verb, 'to look at,' does not convey this meaning, but the German *betrachten,* which is an equivalent, means also to make pregnant ... And if it is pregnant,

> then something is due to come out of it; it is alive, it produces, it multiplies. That is the case with any fantasy image; one concentrates upon it, and then finds that one has great difficulty in keeping the thing quiet, it gets restless, it shifts, something is added, or it multiplies itself; *one fills it with living power and it becomes pregnant* [italics mine].[8]

Through the psychic engagement with a presence within the image, we are now moving with our attraction into the realm of prayer, worship, adoration and the miraculous — the realm of *telestike* or statue-magic, the age-old technique of invoking the gods into the material world.

Telestike

In the ancient Hermetic texts we learn how one of mankind's greatest achievements is the creation of living statues: 'statues ensouled and conscious, filled with spirit and doing great deeds; statues that foreknow the future and predict it by lots, by prophecy, by dreams and by many other means.'[9] Through 'holy and divine mysteries' the priests implanted the souls of daemons into the ritual statues, using a mixture of plants, stones and spices whose 'natural power of divinity' resonated sympathetically with the god.[10] The spiritual presence was kept in the statue by constant sacrifices, hymns and music which conformed to its nature, and in some way spoke to the celebrants as an oracle. The neoplatonist, Proclus, tells us that 'by means of vivifying signs and names they consecrate images and make them living and moving things'; rites not unlike those of concealing relics in statues of saints, to give them a certain holy power.[11] Through these devotional acts the god became present in its own image, which then functioned as a symbol, being simultaneously both matter and living presence. It is the power of the symbol 'that makes the impossible happen' — brings stone to life — but only *through* the perception of the devotee, which has ceased to be limited to the outer appearance or form of the object.[12] When he or she achieved a particular state of contemplative awareness, marked by the ritual of consecration, matter became transfigured and was seen to wake. To those 'outside' the sacred space, no doubt stone remained stone — not everyone would see the weeping virgin or the beckoning god, for the 'miraculous' would be out of the range of their vision.

The writers of the Hermetic texts were talking of rituals and traditions which emerged out of the mysterious depths of ancient Egypt, where

Figure 4: Tutankhamun, photograph by David Finn, from Egyptian Sculpture; Cairo and Luxor *by Edna Russman. (Copyright © 1989, by permission of University of Texas Press)*

one of the words for sculptor meant 'he who keeps alive.'[13] In particular, funeral effigies created in as perfect resemblance as possible to the dead person would be believed to attract his or her *ka* or life force (equivalent, perhaps, to the astral body) and provide it with another home, or container. In the case of the Pharaoh, this procedure would ensure both his immortality and the continuing presence of his power in Egypt (see Figure 4). The *ka* was often given its own representation, such as the *ka* statue of King Hor (Plate 2), which would originally have been painted and decorated with gold leaf.[14] Here the otherworldly presence of the spirit is suggested by the extraordinary eyes, intended to be as life-like as possible, emphasizing the penetrating vision of spiritual sight. We will consider further the power and function of eyes a little later, but emphasize here that for the Egyptians, as for the Hindus today, the divine empowerment of matter was central to religious life, and in their images the gods were given faculties of sense-perception through the combination of technical skill and hieratic ceremony.

In both cultures, such statues were — are — not *art*, as we would call it, but served the supremely functional purpose of housing divine life. Or perhaps art is art precisely *because* it has this unspoken, secret capacity of capturing the life of the world? Ceremonies such as opening the mouth of the statue and bringing it food would ensure the presence of the *ka,* and statues were often placed deep within the tombs of their dead counterparts, unseen by human eyes. But whereas the Egyptian *ka* statues of human beings have eyes to look into another world, in India the images of deities are dressed, bathed, put to bed, given sight and even breath in rites of consecration to effect their incarnation into this world:

> The image, which may be seen, bathed, adorned, touched, and honoured does not stand *between* the worshipper and the Lord, somehow receiving the honour properly due to the Supreme Lord. Rather, because the image is a form of the Supreme Lord, it is precisely the image that facilitates and enhances the close relationship of the worshipper and God and makes possible the deepest outpourings of emotions in worship.[15]

Do we not perform a similar — albeit secularized — ritual, when we surround ourselves with images of the dead, from monumental sculptures to photos on the mantelpiece? We may ask ourselves whether it is not such a 'primitive' belief to sense that the dead live on in some way both through the attractive power of their likeness, acting as a 'bait' to draw their life-force, but also through the quality of our attention and emotional response — our rituals of remembrance — which keep it there. Perhaps it is the responsibility of the living to 'realize' the immortality of the soul in this way, as the power of our nostalgia for their living presence ignites the flame of latent vitality in their images.

The enchantment of Eros

It was from Egyptian influence around the mid-seventh century BCE that the first Greek sculpture emerged, and with it a new desire to create the perfect human form with an idealized beauty that reflected the serenity and perfection of the gods.[16] The early Archaic statues of *kouroi* or youths in the full bloom of beauty and strength surely reflected the human desire to achieve immortality and perfection.[17] These naked figures, believed to be either votive figures on shrines to Apollo, or heroized mortals on grave monuments, were life-size or more, their legs in a walking posture, their mouths slightly open as if breathing, suggestive of an animating force within (see Figure 5).

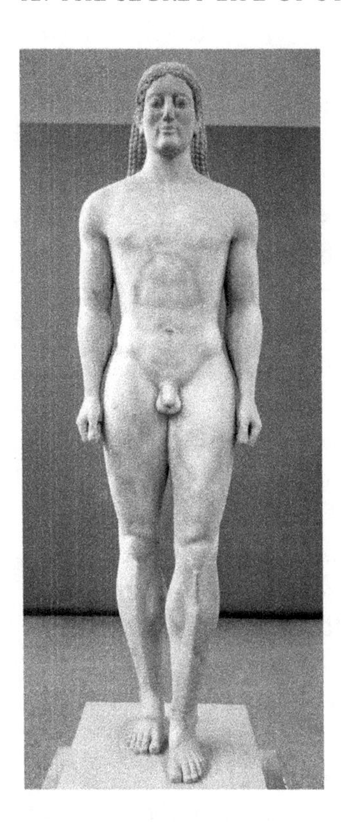

Figure 5: Anavysos Kouros, funerary state statue of Croisus (500–546 BCE), King of Lydia c. *530–520 BC, marble, Greek 6th* c. *BCE, National Archaeological Museum of Greece, Lauros/Giraudon (Bridgeman Art Library)*

For their admirers, they radiated *kallos* and *charis,* beauty and grace, through being what the Greeks called *agalma* – 'an object that through its high quality and craftsmanship inspires delight in its viewer.'[18] But an *agalma* does not only affect the viewer, it also 'prompts' the divinity it imitates to respond by bestowing it with *charis,* the mysterious charm of the gods. Thus divine beings were given material bodies, through which they could meet us half-way and suggest an alluring and enticing other world with their enigmatic gaze– a world which was not impossibly remote, but which gave intimations of accessibility by stirring a long-lost memory. How can we not long to visit the place where such beauty reigns? The equivalent female figures or *korai,* with their elaborate hairstyles and colourful robes radiate *kosmos* or the perfect orderly arrangement of ornament and dress; never naked, like the *kouroi*, their eternal youth and demure self-containment suggest to the viewer that he or she too may transcend the earthly limitations of time and decay.[19] The other-worldly quality intrigues and attracts us, and perhaps it is *this* dimension of beauty with which we fall in love, and to which we aspire in the *cosmetic* enhancement of our bodies. Perhaps, as Deborah Tarn Steiner suggests, 'a work of art does not owe its appeal to its resemblance to a living beloved, but the beloved instigates passion precisely because he or she displays the properties that belong to finely crafted objects' (see Plate 3).[20]

As the archaic style developed into the early classical, the monumental youths and maidens metamorphosed into supple, agile, tactile bodies of such vital presence that they were often bound or chained lest they should escape from their plinths, and were seen to exhibit a variety of movements and gestures.[21] Diodorus Siculus, in the first century BCE, tells us that the mythological sculptor Daedalus:

> … in the production of statues so excelled all other men that later generations preserved a story to the effect that the statues he created were exactly like living beings: for they say that they could see and walk, and preserved so completely the disposition of the entire body that the statue which was produced by art seemed to be a living being. Having been the first to render the eyes open, and legs separately as they are in walking, and also the arms and hands as if stretched out.[22]

The ideal of beauty took the form of the male athlete at the height of his powers, or the *eromenos* or beloved youth whose role in the culture of Greek homoeroticism was to arouse the *eros* of the *erastai*, his older male admirer. But could a boy of flesh and blood ever be a match for the sculptor's *agalma*, whose painted lips, bronzed skin and untouchability were in service to an androgynous beauty beyond the dreams of the most ardent pursuer (see Plate 4)?

The Greeks believed that erotic desire radiated from the beloved, and kindled the longing (*pothos*) of the stricken lover which was not to be appeased, but led on to ever greater intensity. So too with a work of art: 'just like a beautiful person, a beautiful work of art doubly energized the space between it and the observer. Radiating grace (*charis*), it attracted its victim's glance and held it imprisoned.'[23] Unlike a reluctant youth, though, whose combination of reticence and modesty eventually succumbed to the lover's entreaties,[24] a statue denies the physical fulfilment of love forever.[25] If Praxiteles' Aphrodite (Figure 6) ignites love with her erotic glance, as surely she was intended to do, what can the lover do? The famous anecdote of a man who stole into her shrine to make love to the statue[26] graphically demonstrates the irresistible erotic power present in the image — but to judge whether such power emanates from the statue, or from the goddess herself is no easy task. What is certain is that her marble form will not yield to physical assault.

In Praxiteles' hands, for the first time, the female form was deliberately given a sexual allure; look at me, she says, but you can't have me.[27] We may imagine Aphrodite laughing at the clumsy attempt of her devotee to literally consummate his passion — for she knows that true union can only take place in the soul, and that is where her material form beckons. The need to create images of the divine beauty who seduces but denies has taken hold of the human imagination ever since, and in the realism of photography and film the knife-edge of impossible desire becomes even more sharpened and refined (see Figure 7).

Figure 7: Assia by Dora Maar, 1934. From Christian Bouqueret (ed.), Assia sublime modèle *(Mont-le-Marsan 1993), pl.83, in A. Stewart* Art, Desire and the Body *(figure 63)*

Figure 6: Torso of the Venus of Knidos, copy of the original statue of c. 350 BCE (marble), after Praxiteles (c.400–c.330 BCE), Louvre, Paris, France, B. De Sollier & P Muxel (Bridgeman Art Library).

These *are* living, breathing beings we see, and yet they are still beyond our reach — if they weren't, would we still adore them? It would seem that the rite of passage — whether via an ancient goddess or a contemporary film star — however painful, is unavoidable, if one follows the path of love. However we represent our Psyche, waking to her means touching the quick of our own desire for possessing her, and falling to the enchantment of that mighty daemon, Eros.

How different these pale statues would look with their original paint, situated in their temples or shrines, and not alienated in art galleries as 'exhibits.' Even shocking perhaps, in their conflation of divine perfection and human warmth. In Egyptian and Greek traditions, the most significant moment — the moment of animation — was that of the painting in, or inserting, the statue's eyes. In the Hindu ritual of *darsan,* it still is. 'Not only must the gods keep their eyes open' writes Diana Eck, 'but so must we, in order to make contact with them, to reap their blessings, and to know their secrets.'[28] This was the ritual of consecration, the 'making sacred,' the point at which the god entered the image and it became operative, seeing the world as we see it — and meeting our gaze. In Egypt the eyes were often made of polished rock crystal to imitate the human eye with startling accuracy (see Plate 5).[29]

These eyes are far-seeing, penetrating into realms inaccessible to the living, but in Greek statues the otherworldliness becomes part of a *seductive* intent — for they knew that *eros* was contagious through the eyes, and love was born at the sight of beauty.[30] Furthermore, if the statue can see us, then it can fall in love with us, and we are rendered vulnerable, no longer in control, compelled to change our assumptions of superiority over an inanimate object. A channel of communication is opened, because the statue is no longer an image which is separated from life, but one which participates in it.

Take the Apollo from the West Pediment of the Temple of Zeus at Olympia, arresting the fight between Lapiths and Centaurs with his indomitable command and spiritual presence, a monument to the civilizing influence of the gods over man and beast.[31] What happens to our relationship with this deity, when his familiar purity and otherworldliness are bathed in the vibrant colours of this world, when his eyes are restored, and he sees (see Plate 6)?

We can now *recognize* him, for the god has incarnated. In bringing their statues to life through colour and vision, the Greeks created the perfect conditions for the working of *telestike,* where the outer form of the object resonated so harmoniously with its divine model that the soul of the latter was immediately attracted into it, and the two dimensions fused. The neoplatonist Proclus recognized that in theurgic ritual a superbly crafted statue became a receptacle for a mysterious transcendent presence, which he called 'divine illumination.'[32]

Platonic reflections

Which brings us to the neoplatonic understanding of the use of images in the journey of the human soul towards unity with the One, the fount and fulfilment of all being. Plotinus explains how material objects can act as 'baits' or 'lures' to catch the hidden properties of the *anima mundi* — that mysterious life-force which pervades the cosmos — through sympathetic attraction. He explains:

> I think that the wise men of old, who made temples and statues in the wish that the gods should be present to them, looking to the nature of the All, had in mind that the nature of soul is everywhere easy to attract, but that if someone were to construct something sympathetic to it and able to receive a part of it, it would of all things receive soul most easily. That which is sympathetic to it is what imitates it in some way, like a mirror able to catch the reflection of a form.[33]

So, images may be created with the intention of attracting higher, immaterial properties through imitation and reflection; they can thus ultimately become the receptacles of the Divine Ideas themselves, as properties of the Ideas are seeded in the world-soul as it mediates between the divine and human realms. In this way man can make contact with the gods who inhabit the 'middle realm' as messengers, through praise and worship using ritual objects, or movements, music and incantations which conform to their nature. In the *Corpus Hermeticum* Tat says 'there are reflections of the incorporeals in corporeals and of corporeals in incorporeals — from the sensible to the intelligible cosmos, that is, from the intelligible to the sensible. Therefore, my King, adore the statues, because they, too, possess ideas from the intelligible cosmos.'[34]

The rituals of statue animation would have formed part of the theurgy or divinatory ritual of Iamblichus or Proclus. This 'divine work' had, as its ultimate goal, no less than the divinization of the human soul, as the theurgist used symbolic objects and images in order to move to a deeper level of perception not through 'intellectual understanding' or 'discursive thinking' but through ritual acts which awoke a deep *eros*, or sense of primordial participation with divinity. As Gregory Shaw has explained: 'the rituals of theurgy allow us to move from the periphery of embodied awareness to its divine centre. Ultimately, it allows the gods to appear in embodied life, to reveal themselves in human form through our mortal existence.'[35] Iamblichus emphasizes that the connection with divinity does not occur through our intellectual theorizing, but through allowing the power of the symbol to speak to a deeply intuitive sense of *unity* with the divine that can only be activated through ritual.[36] This knowledge is superior and more ancient than any rational judgement, and most importantly, its depths are revealed in proportion to our *desire* to reach it. To quote Shaw again: 'it is not by our knowing, calculating or predicting that we ascend to the gods, but by the intensity and quality of our longing.'[37] In the words of Proclus, the symbolic properties of theurgic images 'move everything towards the desire of the good and this wanting produced in things is unquenchable.'[38]

Platonically speaking, symbols engage us not just intellectually, but also emotionally, inducing *desire* and *longing* for that which they point to, and, crucially, it is the very kindling of that desire which enables us to see them as symbols at all. In other words, symbolic perception is a mobile process which progresses according to the intention, and attention, of the observer. Let us take this a bit further now, and consider what our theurgists meant by moving from a condition of 'separated' knowing to a 'unitary connection with the gods.' Proclus speaks of four different levels in which sense objects participate in divine life, and through which we may respond to them.[39] Firstly, they are quite simply sense-objects, seen literally as just matter. It's only stone, or wood — we might say of a statue — expertly carved, but the bottom line is its materiality. Secondly, a statue may be seen as an image, or *representation*, of something, such as a god. This is an allegorical move, but does not yet bring the image into single focus, as it were, with its archetypal essence, or with the viewer's inner eye — it remains a conceptual exercise. The key move is the next one, where the object is seen as a *reflection* of something beyond it, a symbol which reveals the essence of that which it points to, and must engage us emotionally in order to be grasped (when it sees *us*, perhaps). Finally, the image is seen as *participating* in the very essence of that which it embodies, and embracing us in that essence.[40]

As Proclus points out, the object itself partakes of all these dimensions *simultaneously*, but whether we can follow through to the end depends on *our* ability to move from an outer objectivity to an inner identification; in other words, we bestow it with life through the quality of our response. There can be no empirical truth, no final definition, no categorical statement about what is true or false. In fact 'normal' perception is thrown on its head: for to perceive the outer form as most 'real' is, from this perspective, the most illusory stage of looking, a mere preparation. The most authentic vision is that which sees the inner life as primary, determining the 'appearance' that the imagination perceives.[41] The 'seeing' of the god in the image turns it from an idol into an icon.

Again according to the Platonic tradition, once the senses are caught by the beauty of the image, and desire is kindled, the process of spiritual 'purification' is set in motion — and so we come to the reason, from a metaphysical point of view, *why* the beloved must be unobtainable. One mustn't stay at the level of the purely sensual, but move forward — and how better to do this than through the power of art, for images or statues cannot respond, on the physical level, to our desire. Instead, like the lover in Plato's *Phaedrus*, the soul is led into an impossible position of unrequited passion, a state of extreme tension which forces the growth of its wings, at the same time as leading to an ever clearer vision of divinity and sense of affinity with it.[42] For the Platonists, trusting the intellect rather than the senses impelled the soul to find union on a deeper level, for by the laws of cosmic attraction soul will always seek to unite with itself. In Platonic terms, the lover comes to realize that in the contemplation of beauty he (or she, although for Plato the context is exclusively male) cannot possess, he is being asked to change, to awake to a new and deeper self-consciousness through the exposure of his soul to itself, as in a mirror-image.

We find in Platonism, as in Christianity, the intensely problematic differentiation between the realms of physical and spiritual procreation, and thus between human and divine love. Remaining at the level of the senses — the literal — in both traditions is a danger, yet paradoxically Platonists understood that it was only through sensual, erotic attraction that the journey could begin. For orthodox Christianity the issue has been more difficult to resolve, for dominant schools of thought have removed Eros, like Psyche, entirely from the arena of human sexuality. Images which attract the soul must therefore avoid any hint of sensuality, lest they bind it to the level of the flesh. But as I have already mentioned, it is precisely when there is no channel between matter and spirit that images become idols, devoid of sacred presence, or even worse, are assumed to be receptacles for evil demons. For in the subsuming of all cosmic powers into Christ, the daemons of the pagan cosmos were rendered diabolic, and under such stricture the human imagination could only cease to be an organ of spiritual perception and become regarded as unreliable and deceptive, if not dangerous — to be subjected to Reason as the only legitimate path to divine truth.

We may find an antidote to such uncompromising dualism in Islamic mysticism, with its strong Platonic and Hermetic undercurrents. Here we are given a context in which the imagination is the very ground in which the literal becomes transformed into the spiritual, the place of *theophany* and revelation, the place where Psyche invites us to meet her. We are not talking here of imagination as a kind of 'fantasy' or human invention — which is in fact a mode of distancing ourselves from the 'truly real' — but a source of divine knowledge that flows into the world through human creativity. Despite the prohibition against figural representation in Islam — for the artist, in depicting a living being, was considered to usurp the creative power of God — the emphasis of the Sufi mystics on the transformative power of the Active Imagination is particularly relevant to our consideration of symbolic perception and can be applied to any object of our gaze, inner or outer.[43] The visionary mysticism of Avicenna, for example, emphasizes the function of the imagination as a place of prophetic inspiration; the images which are formed in it do not derive from external perception, but arise in the depths of the soul through the agency of the Angelic hierarchies and are then given material presence in the object of vision.[44] In this way they are reflected back to the soul which transmutes them into symbols.[45]

For sages such as Ibn Arabi, the power of the imagination as the intersection between physical reality and spiritual intellect becomes primary, with an autonomous reality, its own world (called the *mundus imaginalis* by Henry Corbin), and its own organ of cognition.[46] This organ — the *himma* — is a creative power which arises in the heart. Henry Corbin defines *himma* as 'the act of meditating, conceiving, imagining, projecting, ardently desiring.'[47] It is a passionate force which, when directed by the initiated Gnostic, can be even powerful enough to 'manifest' a being externally, to create changes in the outside world. Corbin says: 'thanks to the Active Imagination, the gnostic's heart projects what is reflected in it (that which it mirrors); and the object on which he thus concentrates his creative power, his imaginative meditation, becomes the *apparition* of an outward, extra-psychic reality.'[48] Thus a theophany can occur, a vision of another reality perceivable by those of like consciousness. On one occasion, for example, the Angel Gabriel took the form of a beautiful Arab youth, but only Ibn Arabi saw the angel — his companions saw only the youth.[49] It is as though the image — whether youth or statue — provides the starting point for the *himma* to create its own, internal image which then fuses with, and transfigures, the external one. It is what happens when we fall in love; the object of our desire stirs a deeper vision which then enhances — even divinizes — the beloved, and the whole world is seen through a heightened vision. This double-seeing, a seeing *through*, is instantaneous, as Proclus described. It is an initiation, an awakening. It can happen in the street, or in the art gallery, in reading a poem, on hearing a piece of music. However it happens, however fleeting, it can change lives.

Awakening Psyche

I suggested in the introduction to this paper that our response to images in some way determines the fate of Psyche, in either denying her a place to reside or opening the world to her presence. I have also suggested that she is enticed there through eros, that our desire and longing for union with her is the force which attracts her to the world. In the case of a seductive image, an intense exchange may be built up which does not dissipate, or fulfil itself in consummation, but holds us captive, traps us, engages us, enchants us with a never-ending promise of blissful union, if only we knew the secret of awakening it. That we *can* awaken it is suggested in our myths, fairy tales and sacred texts — which also warn us however that our own desire will not be enough, and that for a miracle to occur, a little supernatural intervention may be required.

We all know the story of Pygmalion and Galatea in Ovid's *Metamorphoses* (see Plate 7); the artist who created a statue of the most beautiful woman, fell in love with her, and, not daring to ask for the statue itself to come to life, appealed to Aphrodite to send him a wife just like her. But Aphrodite did more than that — when Pygmalion returned home from her shrine, he kissed Galatea and she began to awake from her stony slumber and respond to his desire.[50] In Ovid's story the combination of Pygmalion's longing, his invocation to Aphrodite and her response allow a miracle to happen — she 'consecrates' his act of creation by the firing of Eros' arrow and, as the goddess of Love, becomes present in Galatea's image. Pygmalion's intense longing for Galatea was not enough to bring her to life, but through his ritual action the goddess gave a sign that she would intervene on his behalf. The message of this story lies in the surrendering of human will to divine will, the religious act which sanctifies the presence of Psyche and allows her to embody her image. If we want our statues to live, we must first go to the temple.

We find further examples of this theme in the story of story of Eros and Psyche,[51] (Figure 8) the fairytale Sleeping Beauty (Plate 8), and the Christian story of the annunciation of the Virgin Mary (Figure 9). In all these narratives, it is the erotic encounter that transforms and awakes the female soul — 'impregnates' her — through the action of the god (whether as Eros, the Prince or the Angel). In art we see this moment depicted in Botticelli's *Primavera,* as Chastity is fired by Eros' arrow with love for Mercury,[52] and most strikingly in Bernini's scultpure *The Ecstacy of Saint Teresa* (see Figure 10). Here the burning arrow of the angelic Eros induces a mystical union which is, to judge from Teresa's depicted rapture (and her written description of the event), experienced as deeply physical.[53]

Figure 8: Psyche Revived by the Kiss of Love, 1787–93 (marble) by Antonio Canova (1757–1822), Louvre, Paris, France. Giraudon, Italian (Bridgeman Art Library)

Figure 9: Cavalcanti Annunciation (gilded limestone) by Donatello, c.1386–1466 Santa Croce, Florence, Italy (Bridgeman Art Library)

Figure 10: Ecstacy of St Teresa (marble) by Bernini, Giovanni Lorenzo (1598–1680) Santa Maria della Vittoria, Rome (Bridgeman Art Library)

The key to all these awakenings is in the *seeing*, which is initiated by the kiss, Eros' arrow or divine messenger — in other words, the revelation of the god to the human through the intertwining of sensual and spiritual experience. Apollo and Rilke, Galatea and Pygmalion, Psyche and Eros, Beauty and the Prince, Mary and the Angel, Teresa and God, all *recognize* themselves in the other. I would like to suggest that this awakening through eros is available to us too through our contemplation of images, but only if we peel the scales of habit from our eyes, learn how to let go of our defences, and believe in miracles.

Postscript: living statues

The modern imagination is still gripped by the fascination of life in statues. There are many circus and entertainment companies who provide 'living statues' for celebratory events, a mode of entertainment which has its roots in the European theatre tradition of mime and tableaux, but which has now become part of corporate culture and even street life. The advertizing language used by these organizations reveals a need we humans seem to have to witness the animation of created figures, and to be seen by them: 'Ancient white stone figures actually begin to move' says one, 'You draw near, suddenly YOU are the subject of interest.' 'With elegance and grace, they come alive!'[54]

These party acts, in 'literally' bringing statues to life, breaking the taboo as it were, stop our imagination from doing the work and, therefore, from making the deeper connection on the level of the soul. They reduce the divine to the human, the symbolic to the literal, in the attempt to provide a 'quick fix' of fulfilled desire for the audience. We can say 'Oh it's really a person' and the spell is broken, Psyche is again banished, and we don't have to leave our comfort-zone. But perhaps the living statues themselves find that in standing motionless for hours under people's fascinated scrutiny they undergo a kind of initiation, in a re-enactment of a mystery ceremony in which they find out what it is like to be a marble god, brought to life by human attention. In turn, the onlookers may be 'turned to stone' as they gaze in fascination, waiting for a twitching eyelid to betray a sign of life.

Nor is the modern world of advertizing immune to the power of *telestike* as a bait for custom. In 1990 the Italian company Fendi brought out an advertisement for their perfume *La Passione di Roma*, in which a beautiful young woman is seen kissing an ancient statue with an expression of intense longing. Two years later, the same young woman is seen in another advertisement, this time for a perfume called *La Passione Viva*. But now the statue has metamorphosed into a handsome man, at whom she gazes rapturously — although in true Platonic spirit, his eyes do not meet hers, but contemplate, like Botticelli's Mercury, higher realms. The message is loud and clear. If you use the right ritual substance and kiss him passionately enough, you will animate your statue who will then lead you to another world. They could just be right.

Notes

(See bibliography for full references)

1 From *Ahead of all Parting* by Rainer Maria Rilke, translated by Stephen Mitchell, copyright © 1995 Stephen Mitchell. Used by permission of Modern Library, a division of Random House, Inc.
2 Carotenuto, p. 22.
3 Freedberg, p. 17.
4 To quote Freedberg, p. 60: 'But if in the ornament we sense, consciously or unconsciously, the presence of the living (whether human or plant), we cannot so comfortably assign it the status of the decorative, especially in the way in which we speak of the purely or merely decorative. If, in the end, the referent turns out to be a human form, then the fusion of image and prototype troubles the demands we place upon ourselves for aesthetic detachment and disinterested formal perception, confronting us with the disturbing presence of the animate in the inanimate.'
5 Le Guin, at http://www.polyamory.org/~howard/Poetry/rilke_archaic_apollo.html
6 Bruce S. Thornton has explored the destructive and terrifying force of Eros in *Eros: The Myth of Greek Sexuality.*
7 Carr, p. 5.
8 Jung, C. G., *Interpretation of Visions*, privately mimeographed seminar notes of Mary Foote, 1941, Vol.6, Lect.1, May 4, 1932, p. 3; quoted in Chodorow, p. 7.
9 *Asclepius*, p. 24 (Copenhaver trans., *Hermetica*, p. 81)
10 A common practice in antique magical ritual in general, see examples in Betz. Also, Tarn Steiner, p.114.
11 Proclus, *In Timaeum*, 37 c-d, quoted in Freedberg, p. 88.
12 Struck, p. 213. In chaps. 6 and 7, Struck offers a penetrating account of the symbolic use of talismanic objects and statues in the neoplatonic rituals of Iamblichus and Proclus.
13 Gombrich, p. 33: 'The Egyptians held the belief that the preservation of the body was not enough. If the likeness of the king was also preserved, it was doubly sure that he would continue to exist for ever. So they ordered sculptors to chisel the king's head out of hard, imperishable granite, and put it in the tomb

where no one saw it, there to work its spell and to help his soul to keep alive in and through the image.'

14 See Russmann, pp. 75–78.

15 See Eck, pp. 46–55; this quotation, p. 46.

16 On the influence of Egypt on Greek sculpture, see DeFaveri.

17 See http://www.ancient-greece.org/art/kouros.html; Ridgway, pp. 45–83; Stewart (1990), pp. 109–110.

18 Tarn Steiner, 116.

19 See http://www.etymonline.com/index.php?term=cosmos: 'orderly arrangement (cf. Homeric *kosmeo,* used of the act of marshalling troups), with an important secondary sense of 'ornament, decoration, dress.' Pythagoras is said to have been the first to apply this word to 'the universe.' See Ridgway, pp. 85–114.

20 Tarn Steiner, p. 195.

21 Tarn Steiner (pp. 160–1, 174–5, 182) mentions examples in ancient writers of statues changing expression, closing their eyes, speaking, singing, giving oracles, sweating and moaning; they can also 'paralyse, blind and madden men' with their gaze.

22 Diodorus Siculus, 4.76.1–3, quoted in Freedberg, pp. 36f.

23 Stewart (1997), p.19.

24 Tarn Steiner, pp. 205f, Pindar, *Pythian Ode* 9.12.

25 Tarn Steiner (p. 249) remarks: 'Able to arouse *pothos* but not to satisfy the passion it instigates, it emblemizes both the objectification of the *eromenos* through the lover's eyes, and the power, remoteness and ever representational statues with which this transformation endows its original object.'

26 Recounted in Lucian, *Amores,* pp. 13–17; Pliny, *Historia naturalis,* 36, pp. 20f.

27 See Stewart, 1997, p. 104: 'The all-powerful goddess offers [the male spectator] no closure, no safe haven for his desire. Instead, she makes him but one member of a putative love triangle, holding him in her grip like putty, able at her whim either to turn, smile, and bestow unimaginable bliss upon either him or his rival, or avenge their trespass with devastating effect.' Tarn Steiner remarks (p. 219) that by the early fifth century sculptors were using nudity 'to invest the body with a markedly sexual edge.'

28 Eck, 1998.

29 See Russmann, p. 18.

30 See Frontisi-Ducroix; Elsner.

31 See Ashmole and Yalouris, p. 17.

32 Proclus, *In Platonis Cratylum commentaria* 18.27–19.18: 'By [its representational power] the soul is able to assimilate itself back up to its superiors — gods, angels, and daemons. But further, by the same power it also likens the things that descend from itself to itself and, further still, to those things superior to itself. On account of this it fashions statues of both gods and daemons. But wishing to bring into being similitudes of the things that exist that were in a certain way immaterial and born from intellectual substance only, .. it brought forth from itself the substance of names. And just as the ritual art in this way makes statues like gods through certain symbols and unspeakable *synthemata*

(symbols) and makes them suitable for divine illuminations, so too by the same power of assimilation the lawgiver's art brings into being names as statues of their objects …' (quoted in Struck, p. 236).

33 Plotinus, *Ennead* IV.3.11

34 *Corpus Hermeticum* 17 (Copenhaver, trasl., p. 62)

35 Shaw, 'Astrology as Divination: Iamblichean theory and its contemporary practice,' p. 5.

36 Iamblichus, p. 96: 'Intellectual understanding does not connect theurgists with divine beings, for what would prevent those who philosophize intellectually from having theurgic union with the Gods? But this is not true; rather, it is the perfect accomplishment of ineffable acts, religiously performed and beyond all understanding, and it is the power of ineffable symbols comprehended by the gods alone, that establishes theurgical union. Thus we don't perform these acts intellectually; for then their energy would be intellectual and would depend on us, neither of which is true. In fact, these very symbols, by themselves, perform their own work, and the ineffable power of the Gods with which these symbols are charged, itself, recognizes, by itself, its own images. It is not awakened to this by our thinking.' (Shaw, trans., *Theurgy and the Soul: The Neoplatonism of Iamblichus*, p. 84).

37 Shaw, 'Astrology as Divination: Iamblichean theory and its contemporary practice,' p. 6.

38 Proclus, *In Cratylem* 30.19–32.3, quoted in Struck, p. 237.

39 Proclus (1987), p. 847: 'Sense objects participate in forms by way of impression, but only in the forms presented by nature, for these reason-principles act somewhat like seals when they descend into things. And they receive reflections of the Forms, but only of the forms on the level of Soul, so that they become semblances of souls animated by them with clearer living powers. And they are likenesses when they are made to resemble intellectual Forms, as in the *Timaeus* the visible animal is said to be a likeness of the intelligible. Hence, [a very ingenious person] says, things are likenesses of intellectual Forms, reflections of the soul-forms, and imprints of the physical forms' (Morrow and Dillon, trans.).

40 The transition from image to symbol is explained thus by Struck (p. 221): 'The image marks the material world in its status as a fainter reproduction of a higher principle, but the world seen as *symbol* indicates its status as a manifestation — that is, something that works according to the logic of the trace, with the capacity to point us back up to the higher orders that produced it.'

41 This debate raged in antiquity as now. Lucian attempted to discredit the 'inner life' of statues by describing their 'literal' interiors as messy and cluttered (Tarn Steiner, p. 187); common to philosophical understanding was the assumption that all men see the exterior of things, but only the philosopher can grasp the true essence.

42 The Platonic path of 'taming' the sexual passion of the body in order to strengthen the 'rational soul' is not without its critics, and can certainly be seen as a male imperative in the history of Western attitudes towards sex, and one which devalues the realm of the sensual, nature and the body. For example, Thornton, p. 212: 'we find in Plato's homoerotic metaphysics the most extreme

instance of the mind's attempt to control the body's most volatile appetite by appropriating and exploiting its energy to fuel the transcendence of this messy, chaotic world of change, suffering — and sex.'

43 See Arnold, pp. 5–7.

44 See Corbin (1988), pp. 7–8: 'The Image in question is not one that results from some previous external perception, it is an Image that precedes all perception, an *a priori* expressing the deepest being of the person.. Each of us carries within himself an Image of his own world, his *imago mundi*, and projects it into a more or less coherent universe, which becomes the stage on which his destiny is played out.'

45 See Corbin (1976, 1990), p. 11: 'The active imagination ... will not produce some arbitrary, even lyrical, construction standing between us and 'reality,' but will, on the contrary, function directly as a faculty and organ of knowledge just as real as — if not more real then — the sense organs. However, it will perceive in the manner proper to it: the organ is not a sensory faculty but an *archetype-image* that it possessed from the beginning: it is not something derived from any outer perception. And the property of this Image will be precisely that of effecting the transmutation of sensory data, their resolution into the purity of the subtle world, in order to restore them as symbols to be deciphered, the 'key' being imprinted on the soul itself.'

46 See Corbin (1976).

47 Corbin, (1976, 1990), p. 222.

48 Corbin, *ibid.* p. 223.

49 Corbin, *ibid.* p. 219.

50 In the words of Ovid (*Metamorphoses,* Book 10): 'he went directly to his image-maid, bent over her, and kissed her many times, while she was on her couch; and as he kissed, she seemed to gather some warmth from his lips. Again he kissed her; and he felt her breast; the ivory seemed to soften at the touch, and its firm texture yielded to his hand, as honey-wax of Mount Hymettus turns to many shapes when handled in the sun, and surely softens from each gentle touch. He is amazed but stands rejoicing in his doubt; while fearful there is some mistake, again and yet again, gives trial to his hopes by touching with his hand. It must be flesh! The veins pulsate beneath the careful test of his directed finger. Then indeed, the astonished hero poured out lavish thanks to Venus; pressing with his raptured lips his statue's lips. Now real, true to life — the maiden felt the kisses given her and blushing, lifted up her timid eyes, so that she saw the light and sky above, as well as her rapt lover.'

51 In Apuleius, *The Golden Ass*; see translation and commentary by Neumann.

52 As an illustration of the Platonic function of love, in this painting Mercury does not respond to Chastity but looks up to the heavens and pierces the clouds with his caduceus, implying that love for an individual must be directed towards love for the Good.

53 See St Teresa of Avila, ch. 29, para. 17: 'I saw in [the angel's] hand a long spear of gold, at at the iron's point there seemed to be a little fire. He appeared to me to be thrusting it at times into my heart, and to pierce my very entrails; when he drew it out, he seemed to draw them out also, and to leave me all on fire with a great love of God.'

54 World Gate Entertainment, *The Living Statues,* at http://www.worldgateentertainment.com/The%20Living%20Statues.htm

Bibliography

Apuleius, *The Golden Ass,* P. G. Walsh, trans. (Oxford: Oxford University Press, 1999).

Andronicos, Manolis, Manolis Chatzidakis and Vassos Karageorghis (eds.) *The Greek Museums* (Athens: Ekdotike Athenon, 1975).

Arnold, T. W., *Painting in Islam* (Oxford, 1928 [New York: Dover, 1965]).

Ashmole, Bernard, and Nicholas Yalouris *Olympia: The Sculptures of the Temple of Zeus* (London: Phaidon Press, 1967).

Belting, H., *Likeness and Presence: A History of the Image before the Era of Art* (Chicago: University of Chicago Press, 1994).

Betz, Hans Deiter (ed.), *The Greek Magical Papyri* (Chicago: University of Chicago Press, 1986).

Blackman, A. M., 'The Rite of Opening the Mouth in Ancient Egypt and Babylonia,' *Journal of Egyptian Archaeology,* Vol. 10.

Boardman, John and Eugenio La Rocca *Eros in Greece* (London: John Murray, 1978).

Buitron-Oliver, Diana, *The Greek Miracle: Classical Sculpture from the Dawn of Democracy* (Baltimore: National Gallery of Art, Washington, 1992).

Carotenuto, Aldo, *Eros and Pathos* (Toronto: Inner City Books, 1989).

Carr, David, *The Erotic Word: Sexuality, Spirituality and the Bible* (Oxford: Oxford University Press, 2003).

Chodorow, Joan (ed.), *Jung on Active Imagination* (London: Routledge, 1997).

Copenhaver, B., *Hermetica* (Cambridge: Cambridge University Press, 1989).

Corbin, Henry, *Alone with the Alone: creative imagination in the Sufism of Ibn'Arabi* (Princeton: Princeton University Press, 1969 [1997]).

Corbin, Henry, *Avicenna and the Visionary Recital* (Princeton: Princeton University Press, 1988).

Corbin, Henry, *Mundus imaginalis or the Imaginary and the Imaginal* (Ipswich: Golgonooza Press, 1976) at http://www.geocities.com/Heartland/Woods/4623/mundus.htm

Corbin, Henry, *Spiritual Body, Celestial Earth* (Princeton: Princeton University Press, 1976 [1990]).

DeFaveri, Jonathan, 'Greek Sculpture: The Ideal Human Form' at http://www.bu.edu/uni/iass/hellenism/Greek%20Sculpture.pdf

Eck, Diana L., *Darsan, Seeing the Divine Image in India* (Columbia: Columbia University Press, 1998).

Elsner, John, 'Naturalism and the Erotics of the Gaze,' in N. B. Kampen, *Sexuality in Ancient Art* (Cambridge: Cambridge University Press, 1996), pp. 247–61.

Freedberg, David, *The Power of Images: Studies in the History and Theory of Response* (Chicago: University of Chicago Press, 1989).

Frontisi-Ducroix, F., 'Eros, Desire and the Gaze,' in N. B. Kampen, *Sexuality in Ancient Art* (Cambridge: Cambridge University Press, 1996), pp. 81–100.

Gombrich, Ernst, *The Story of Art* (Oxford: Phaidon Press, 1972).

Gross, Kenneth, *The Dream of the Moving Statue* (Ithaca: Cornell, 1992).

Halperin, Winkler and Zeitlin, (eds.), *Before Sexuality: The Construction of Erotic Experience in the Ancient Greek World* (Princeton: Princeton University Press, 1990).

Hanegraaff, Wouter J., 'Sympathy or the Devil? Renaissance Magic and the Ambivalence of Idols' at http://www.esoteric.msu.edu/VolumeII/Sympdevil.html

Iamblichus, *De mysteriis,* in trans. Emma C. Clarke, John M. Dillon, Jackson P. Hershbell, *Iamblichus: On the Mysteries* (Atlanta: Society of Biblical Literature, 2003).

The Way of Hermes trans. C. Salaman, D. van Oyen, W. Wharton (London: Duckworth, 1999)

Kampen, N. B. (ed.), *Sexuality in Ancient Art* (Cambridge: Cambridge University Press, 1996).

Koloski-Ostrow, A., and C. L. Lyons, (eds.), *Naked Truths: Women, Sexuality and Gender in Classical Art and Archeology* (London and New York: Routledge, 1997).

LeGuin, Ursula, 'Myth and Archetype in Science Fiction,' in *The Language of the Night, Essays on Fantasy and Science Fiction* (New York: G. P. Putnam's Sons, 1979).

Mitchell Havelock, Christine, *The Aphrodite of Knidos and her Successors: A Historical Review of the Female Nude in Greek Art* (Michigan: University of Michigan Press, 1995).

Mitchell, Stephen, trans., *Ahead of All Parting: Selected Poetry and Prose of Rainer Maria Rilke* (New York: Modern Library, 1995).

Neumann, Erich, *Amor and Psyche: The Psychic Development of the Feminine, a Commentary on a Tale by Apuleius* (Princeton: Princeton University Press, 1971).

Ovid, *Metamorphoses,* A. D. Melville, trans. (Oxford: Oxford University Press, 1998).

Plotinus, *Enneads,* IV, Stephen McKenna, trans. (London, 1926, repr. Harmondsworth: Penguin Books, 1991).

Proclus' Commentary on Plato's Parmenides, trans. G. Morrow & J. Dillon (Princeton: Princeton University Press, 1987).

Proclus, *On the sacred heart (De sacrificio et magia),* trans. S. Ronan, at http://www.esotericism.co.uk/proclus-sacred.htm.

Rappe, Sara, *Reading Neoplatonism: Non-discursive thinking in the texts of Plotinus, Proclus and Damascius* (Cambridge: Cambridge University Press, 2000).

Ridgway, B. S., *The Archaic Style in Greek Sculpture* (Princeton: Princeton University Press, 1977).

Russmann, E. E. (ed.), *Egyptian Sculpture, Cairo and Luxor* (London: British Museum Publications, 1990).

Shaw, Gregory, 'Astrology as Divination: Iamblichean theory and its contemporary practice' (unpublished paper, 2005).

Shaw, Gregory, *Theurgy and the Soul: The Neoplatonism of Iamblichus* (Pennsylvania: Pennsylvania State Press, 1995).

Stewart, Andrew, *Art, Desire and the Body in Ancient Greece* (Cambridge: Cambridge University Press, 1997).

Stewart, Andrew, *Greek Sculpture* (Newhaven and London: Yale University Press, 1990).

Struck, Peter, *Birth of the Symbol: Ancient Writers at the Limits of their Texts* (Princeton: Princeton University Press, 2004).

St Teresa of Avila, *The Life of St Teresa of Avila* trans. David Lewis (London: Burns and Oates, 1962).

Tarn Steiner, Deborah, *Images in Mind: Statues in Archaic and Classical Greek Literature and Thought* (Princeton: Princeton University Press, 2001).

Thornton, Bruce S., *Eros: The Myth of Ancient Greek Sexuality* (Boulder, Colorado: Westview Press, 1997).

The Sophia Centre

The Sophia Centre is an academic body within the School of Historical and Cultural Studies at Bath Spa University. At present the department offers a Masters programme in Cultural Astronomy and Astrology. The department also supervises MPhil/PhD research and offers BA modules within other disciplines at Bath Spa University.

The Centre studies the impact of astrological and astronomical beliefs on cultures, religions, politics and the arts. Beliefs about the influence of the stars upon human affairs have affected the political, cultural and religious development of different societies, from the past to the present day. Whilst the scientific study of astronomy is well established in universities, the academic study of astronomy and astrology from a cultural perspective has not been the specific focus of western university research for over three hundred years.

For enquiries and applications, please contact Dr Nicholas Campion: ncampion@caol.demon.co.uk, or see the Sophia Project website at: http://www.sophia-project.org.uk/.

The Sophia Centre was set up with financial support from the Sophia Trust, a charity established for advancing education through teaching the relationship between knowledge and beliefs about the heavens and every aspect of art, science and philosophy.

Contributors

Bernadette Brady has a MA in Cultural Astronomy and Astrology and is currently involved in PhD studies at the Sophia Centre, Bath Spa University. She is author of *The Eagle and the Lark: a Textbook of Predictive Astrology* (Weiser, 1992 and 1998) and *Brady's Book of Fixed Stars* (Weiser, 1998).

Professor Neville Brown is a Senior Member of Mansfield College, Oxford University. He served (1957–60) as a meteorologist with the Fleet Air Arm. Subsequently specializing in defence studies, he became a professor at Birmingham University in 1980. He was an accredited Defence Correspondent (1966–73) in the Middle East and Southeast Asia. At Mansfield, he served (1994–97) as the academic consultant to an MOD task force reviewing policy towards Ballistic Missile Defence. Generally, his interests have evolved more towards sky science. His books include *History and Climate Change* (Routledge, 2001) and *Engaging the Cosmos: Astronomy, Philosophy and Faith* (Sussex Academic Press, 2006).

Dr Nicholas Campion is Principal Lecturer in History and Director of the Sophia Centre at Bath Spa University. He is course director of the MA in Cultural Astronomy and Astrology and editor of *Culture and Cosmos: A Journal of the History of Astrology and Cultural Astronomy*. His books include *What do Astrologers Believe?* (Granta, 2006), *Cosmos: A Cultural History of Astrology* (London and Hambledon, 2007) and *A Historical Dictionary of Astrology* (Scarecrow, 2008).

Jules Cashford is the author of *The Moon: Myth and Image*, and co-author, with Anne Baring, of *The Myth of the Goddess: Evolution of an Image*. She has a background in Philosophy and Literature, is a member of the International Association of Analytical Psychology, and lectures on Mythology. She has translated *The Homeric Hymns* for Penguin Classics.

Noel Cobb is a Fellow of the Temenos Academy and an accredited (UKCP) psychotherapist. In 1988, together with the late Eva Loewe, he founded the *London Convivium for Archetypal Studies*, and the annual publication, *Sphinx: A Journal for Archetypal Psychology and the Arts*. He has published seven books of poetry and translated (with Eva Loewe) Rilke, Trakl, Neruda and Lorca. He is author of *Prospero's Island: the Secret Alchemy at the Heart of The Tempest* and *Archetypal Imagination: Glimpses of the Gods in Life and Art.*

Dr Patrick Curry has a PhD in the History and Philosophy of Science from University College, London (1987). He is Lecturer in Religious Studies at the University of Canterbury, Kent. He is currently the assistant editor for the journal *Culture and Cosmos*, and his most recent publications are *Astrology, Science and Culture: Pulling Down the Moon* (Berg, 2005), co-authored with Professor Roy Willis, and *Ecological Ethics: An Introduction* (Polity Press, 2006).

Cherry Gilchrist is the author of numerous books including *Elements of Alchemy, The Circle of Nine,* and *A Calendar of Festivals*. She has made a personal study of the Western esoteric tradition for the last thirty-five years, and for the last twelve has also been studying Russian traditional life, art and culture. She lectures widely on these subjects, and lives and works in Bath. She is currently completing a MA in Cultural Astronomy and Astrology at Bath Spa University.

Dr Liz Greene is director of the London-based Centre for Psychological Astrology (CPA) and the CPA Press, and a visiting lecturer on the MA in Cultural Astronomy and Astrology at Bath Spa University. A professional astrologer and Jungian analyst, she is the author of numerous pioneering works on astrological, psychological and mythological themes, including *Saturn: a New Look at the Old Devil, Relating, The Astrological Neptune and the Quest for Redemption* and most recently *The Dark of the Soul.*

Robert Hand is an author of numerous seminal works in astrology, including the best-selling *Planets in Transit*, *Planets in Youth*, *Planets in Composite*, and *Horoscope Symbols*. Renowned for his pioneering work into new astrological techniques and research, he is the founder of the Archive for the Retrieval of Historical Astrological Texts (ARHAT), and is currently involved in Project Hindsight as an editor, translator and publisher of ancient astrological writings. Rob lectures all over the world and lives in Virginia, USA.

Dr James Hillman has achieved legendary status in his own lifetime and is one of the most inventive and original psychologists in the world. A Jungian analyst and the originator of post-Jungian 'archetypal psychology,' he is the author of more than forty books including *The Force of Character and the Lasting Life, Re-Visioning Psychology, Healing Fiction, The Dream and the Underworlds, The Myth of Analysis* and *Suicide and the Soul.* His most famous book, *The Soul's Code,* was the top title on the *New York Times* bestseller list.

Dr Jarita C. Holbrook, has a BSc in Physics from Caltech, an MSc in Astronomy from San Diego State University and a PhD in Astronomy and Astrophysics from UC Santa Cruz. After receiving her PhD in 1997, she went to Africa to begin work in cultural astronomy in Tunisia. Since then she has dedicated herself to expanding the number of African cultural astronomy scholars and research in this field. She is currently initiating a project to study the links between astronomy and ritual dance in Africa.

Sir Nicholas Pearson was educated at Radley College, Oxford, after which he joined the Army. As a soldier he saw active service in Cyprus, Borneo and Central Africa. He has been a Director of Virgin Atlantic Airlines and a Cultural and Business Advisor to one of the great dynastic families of Japan. He has served as a local District Councillor and was the Conservative Parliamentary Candidate for Oldham, Manchester, in the 1970s. He is currently Chairman of the Temenos Academy, and has recently trained as a psychotherapist.

Dr Richard Tarnas is professor of philosophy and psychology at the California Institute of Integral Studies in San Francisco, where he teaches archetypal studies and the history of Western thought and culture. He was the founding director of the PhD and Master's programme in Philosophy, Cosmology, and Consciousness. He also teaches on the faculty of the Pacifica Graduate Institute. He is author of *The Passion of the Western Mind*, a narrative history of the Western world view from the ancient Greek to the postmodern. His most recent book is *Cosmos and Psyche: Intimations of a New World View* (Viking, 2006).

Dr Angela Voss is currently convenor of the Cosmology and Divination course for the MA in the Study of Mysticism and Religious Experience at the University of Kent. She is also a musician and an astrologer, with interests in Renaissance magic and astrology, music performance practice, the relationship of the imagination to spiritual experience and the religious function of images and symbols. In addition to many published papers on Marsilio Ficino and Renaissance esotericism, she has produced two recordings, *Secrets of the Heavens* and *Images of Melancholy* (Riverrun Records).

www.ingramcontent.com/pod-product-compliance
Lightning Source LLC
Jackson TN
JSHW011352130125
77033JS00016B/567

* 9 7 8 0 8 6 3 1 5 5 6 6 6 *